EXPERIMENTAL MICROBIOLOGY

EXPERIMENTAL MICROBIOLOGY

F.C. GARG
Professor of Microbiology,
CCS Haryana Agricultural University,
Hisar - 125 004 (India)

CBS

CBS Publishers & Distributors Pvt. Ltd.

New Delhi • Bengaluru • Chennai • Kochi • Kolkata • Mumbai
Hyderabad • Nagpur • Patna • Pune • Vijayawada

ISBN: 81-239-0938-1

First Edition: 2003
Reprint: 2005, 2009, 2017

Published by **Satish Kumar Jain** and produced by **Varun Jain** for
CBS Publishers & Distributors Pvt. Ltd.,
4819/XI Prahlad Street, 24 Ansari Road, Daryaganj, New Delhi - 110002
delhi@cbspd.com, cbspubs@airtelmail.in • www.cbspd.com
Ph.: 23289259, 23266861, 23266867 • Fax: 011-23243014

Corporate Office: 204 FIE, Industrial Area, Patparganj, Delhi - 110 092
Ph: 49344934 • Fax: 011-49344935
E-mail: publishing@cbspd.com • publicity@cbspd.com

Branches:

- ***Bengaluru:*** 2975, 17th Cross, K.R. Road, Bansankari 2nd Stage, Bengaluru - 70 • Ph: +91-80-26771678/79 • Fax: +91-80-26771680 E-mail: cbsbng@gmail.com, bangalore@cbspd.com
- ***Chennai:*** No. 7, Subbaraya Street, Shenoy Nagar, Chennai - 600030 Ph: +91-44-26681266, 26680620 • Fax: +91-44-42032115 E-mail: chennai@cbspd.com
- ***Kochi:*** Ashana House, 39/1904, A.M. Thomas Road, Valanjambalam, Ernakulum, Kochi • Ph: +91-484-4059061-65 Fax: +91-484-4059065 • E-mail: cochin@cbspd.com
- ***Kolkata:*** 6-B, Ground Floor, Rameshwar Shaw Road, Kolkata - 700014 Ph: +91-33-22891126/7/8 • E-mail: kolkata@cbspd.com
- ***Mumbai:*** 83-C, Dr. E. Moses Road, Worli, Mumbai - 400018 Ph: +91-9833017933, 022-24902340/41 • E-mail: mumbai@cbspd.com

Representatives:

- Hyderabad: 0-9885175004
- Nagpur: 0-9021734563
- Patna: 0-9334159340
- Pune: 0-9623451994
- Vijayawada: 0-9000660880

Printed at:
J.S. Offset Printers, Delhi (India)

Dedicated to
My Beloved Mother
Smt. (Late) Parsanni Devi

Foreword

Microbiology has emerged as a fundamental science to explore the mechanisms of biochemical processes involved in cellular metabolism, as it is easier to handle and work with microorganisms that possess common basic pathways. The importance of microorganisms and science of microbiology has been realised only recently. Thus, the subject has also been introduced in Degree Colleges in Haryana and other parts of the country.

I am immensely pleased to learn that Professor F.C. Garg of the Department of Microbiology has written a book entitled 'Experimental Microbiology'. Dr. Garg is one of the teachers having considerable experience in teaching graduate as well post-graduate classes in the subject of microbiology. The book consists of 56 exercises segregated into eleven sections. It is precise and simple to understand. The illustrations add to the usefulness of the book. The publication will fulfill the long felt need of a suitable and well-planned book on experiments in microbiology. I am sure that this will prove to be useful both to students and the teachers. Even the beginners will find the book enticing and easy-to-understand.

I congratulate Professor Garg on this accomplishment.

VINAY KUMAR
Vice-Chancellor,
Chaudhary Charan Singh Haryana Agricultural
University, Hisar - 125 004 (India)

Preface

Microbiology is the backbone of sciences like Biotechnology, Genetic Engineering, Molecular Biology, Biochemistry and Genetics, for microorganisms are not only easy to handle but can also be manipulated at will. However, to understand the principles underlying, one must have the basic knowledge of techniques employed in the study of microorganisms. The importance of microbiology has been realised recently and the subject apart from the universities is also being extended to colleges at graduate level. Moreover, microbiology is one of the important subjects for the students of Biotechnology. Today, both Microbiology and Biotechnology are taught in several institutes all over the country. Both the students and teachers have been facing problems in accomplishing laboratory exercises within the stipulated hours because of paucity of quick microbiological recipes. Thus the book "Experimental Microbiology" has been produced keeping in view the background of the students who, in general, are not much aware of the microbes and their consequential effects.

The book includes 56 exercises grouped into eleven sections, namely Introduction, Control of microbial growth, Microscopy, Microorganisms—An overview, Staining techniques, Culture techniques, Physiological characteristics, The bacterial growth, Microbiology of food, Microbes and environment, and Bacterial genetics. Each section has been introduced with a background information to help students understand the basics underlying each exercise. The book has been prepared to acquaint the students with not only the microbiological techniques but also to make them aware of the importance of microbes in our daily life and their central role in nature. Each exercise has been written in a manner so as to minimize the time required for instructions at the beginning of each laboratory exercise and make more time available for "Do-it-yourself". The illustrations provided in the text wherever necessary add to the usefulness of the book. Each section contains review questions to help the students put themselves to test their comprehension. I believe that the book shall prove to be useful to teachers and students alike in planning and execution of laboratory exercises.

F.C. GARG

Acknowledgements

I am thankful to Sh. Vinay Kumar, Vice-Chancellor, CCS Haryana Agricultural University, Hisar for adding the "Foreword" and permitting me to publish the book. I thank Prof. Dalel Singh, Head, Department of Microbiology, CCS Haryana Agricultural University, Hisar for encouragement and valuable suggestions. I had constantly been encouraged by my colleagues to write a book on practical microbiology. I am indebted to them in general and to Prof. S.S. Dhamija, in particular, who reviewed the manuscript critically and made valuable amendments that indeed helped improve the quality of presentation.

My gratitude goes to my beloved wife Mrs. Usha Rani and our children for their endurance, cooperation and moral support during the preparation of this book.

I appreciate the help of all those, whose names I might have missed, however, unknowingly.

I am thankful to the staff of CBS Publishers & Distributors for timely publication of the book.

F.C. GARG

Contents

Chapter 1

Introduction

Microbiology is the study of living organisms that can not be seen with naked eye without the help of a powerful compound microscope. Thus while studying microbiology we generally are working with unseen living forms of life. Microorganisms are present all around us and play a vital role in ecology of earth. Some are commercially used in the production of industrially important chemicals such as alcohol, organic acids and antibiotics. Microorganisms are also responsible for food fermentations thus enhancing their quality and shelf life. Microorganisms do cause disease in animals including humans and plants and spoil food but the beneficial role they play in our daily life camouflage their harmful effects. Since the techniques employed in microbiology may be new to many of you, you ought to develop perfection. The expertise in these techniques will determine not only the success or failure in your scientific probing but these will be essential to protect you and others around you against potential hazards. You may or may not be working with pathogenic microorganisms but it is essential to observe certain protective measures while working in a microbiological laboratory. Some hints given below shall be observed faithfully for the sake of safety.

1. Students should come in the laboratory with clean laboratory aprons. Aprons should be sterilised in an autoclave before sending for wash in the laundry.
2. Do not bring/leave valuables in your purses.
3. Food must not be eaten in the laboratory.
4. Do not use laboratory ware for drinking of water, tea etc.
5. Keep papers, pencils, fingers and other objects out of the mouth.
6. Observe all possible cleanliness and neatness in the care of apparatus, microscope and desks.
7. Whenever a living culture is spilled, bring to the notice of instructor incharge. Cover it with some disinfectant.
8. All non-infectious solid wastes, such as cotton, paper and matches etc. should be placed in the dust-bin provided for the purpose.

9. All the cultures should be autoclaved before discarding.
10. The inoculation needle used in transferring culture, should be sterilised over the flame before and after use. Flame the wire vertically not horizontally.
11. In case of personal accident such as cutting or pricking of the finger or splashing of culture material in the eye, report immediately to the instructor.
12. At the end, put away all apparatus and clean the desk top with disinfectant.
13. Wash your hands thoroughly with soap and water.
14. Before leaving the laboratory, check that the gas and the water taps are turned off.
15. Keep a record of each day's work, with drawings in a laboratory notebook.

CLEANING OF GLASSWARE

New Glassware: Prepare 1% solution of a good glass cleaner such as teepol/liquid detergent. Place new glassware like test tubes, Petri dishes, pipettes and flasks into the solution and heat to boil. Remove the glassware and rinse thoroughly with warm tap water. Allow to dry before use.

New Microscope Slides: Immerse new slides in alcohol containing hydrochloric acid (3%) for several hours. Remove the slides and rinse thoroughly with tap water and finally in distilled water. Wipe the slides dry with soft cotton cloth. While wiping hold the slides by their edges to prevent recontamination with grease. Keep the slides in a covered jar till use.

New Cover Slips: Cover slips are also cleaned free of grease by the method described for the new microscope slides.

Used Glassware: The used glassware and containers should be discarded in 3% lysol or any other disinfectant. All the glassware must be sterilised in an autoclave prior to washing. This must be followed regardless of whether the glassware is contaminated with pathogenic or saprophytic organisms. After sterilisation, empty the glassware of its contents. Put melted agar into dustbin provided for the purpose. Do not put melted agar into the sink as it will solidify on cooling and shall block the drainpipes. Now wash the glassware like new glassware described earlier.

For biochemical work immerse the washed glassware in dichromate-sulphuric acid solution overnight, remove, wash and rinse in hot water for 3 to 4 times.

Pipettes: Place the used pipettes in 3% lysol solution. If necessary keep immersed in chromic acid overnight, wash first with tap water and than with distilled or deionized water.

Used Slides and Cover Slips: Place slides and cover slips in a pan containing 1% solution of a good cleaner. Heat till boiling and keep for a few minutes. Remove the material, rinse thoroughly in warm tap water and finally in distilled water. Wipe dry with a soft clean cloth using the same precautions as given under new slides and cover slips. Store in a covered jar.

Preparation of Chromic Acid (Cleaning solution)

Dissolve 63.0 g Potassium dichromate by heating in 350 ml distilled water, cool and add concentrated H_2SO_4 into it to make the volume to 1000 ml. Sulphuric acid should be added along the side of the flask slowly. Handle the solution carefully and use rubber gloves and apron.

SPECIFIC HAZARDS

Handling of Alcohol: Both ethyl as well as methyl alcohol is highly inflammable. Keep containers of alcohol covered and away from open flame.

Breakage of Glassware: In case a glass object breaks, do not pick the pieces of glass with bare hands. Sweep the pieces with a broom and dustpan and put into the containers marked for this purpose. However be careful if a thermometer breaks and report to the instructor.

Electrical Equipments: Ensure that your hands are dry when you handle electrical connectors. Don't allow water or any other liquid to come in contact with electrical cords and conductors.

Fire:

1. If you smell a leakage of gas, open the doors and windows, check all gas taps and tighten. Don't switch on any electrical equipment or burn a matchstick until the room is free of the gas.
2. If gas burns from a leak, turn off the gas tap.
3. In case of minor fire, smoother it with a towel or book.
4. If a larger fire occurs, use fire extinguishers in the laboratory to put it out.
5. In cases of large fire involving the laboratory itself, evacuate the room after turning all gas burners off and unplugging the electrical equipment.

Accidents:

1. Report all accidents to your instructor immediately.
2. For spills in or near eyes, use eyewash.
3. For larger spills on the body, use shower.
4. For heat burns, chill the affected part with ice immediately.
5. In case of cut or abrasion, use bandage.

Chapter 2

Control of Microbial Growth

Control of microbial growth is necessary for two reasons:

1. To limit the destruction of a valuable nutrient source i.e. food in food and dairy industry.
2. Uncontrolled growth in animal tissue causes cell destruction, a process well known as infection.

Thus to save the food industry and to prevent the spread of infectious diseases control of microbial growth is must. In general, control can be affected either through inhibition or killing or removal of all viable organisms, a process known as sterilisation.

Microbiological study with pure cultures is impossible if the glassware and culture media were contaminated with any other kind of organisms at the time of use. Thus the test tubes, flasks, pipettes, Petri plates, culture media etc. must be sterilised before use. **Sterilisation** is the process that makes the materials/objects free from all living organisms. Once sterilised, media or glassware must be maintained in sterile condition indefinitely by protecting them from atmospheric contamination. This is ensured by plugging the test tubes, flasks, bottles etc. with cotton and placing the pipettes, Petri plates etc. in aluminium/stainless steel/copper containers before sterilisation. This will protect these materials from dust and outside contamination after sterilisation. A complete destruction of organisms can be achieved by exposure to lethal physical or chemical agents. Solutions of vitamins, amino acids and other growth factors, which are destroyed by drastic heat treatments, are sterilised by filteration. Various methods used for sterilisation are listed in Table 2.1.

A. HEAT STERILISATION

The most commonly used form of physical sterilisation is heat. The killing of microorganisms by heat is an age-old process. Spallanzani (1799) and Pasteur (1860) used heat to destroy microorganisms present in the infusions. The heat used may be dry heat as in case of hot air oven or wet heat as in an autoclave. Moist heat kill the organisms by denaturing proteins (enzymes and structural proteins), while dry heat kill organisms mainly by the oxidation of

Table 2.1. Methods used to destroy microorganisms

Physical					Chemical
Heat		Radiation		Filtration	
Dry heat	Moist heat	Ionizing	Non-ionizing		
Hot air Infra red	Steam under pressure Tyndalization Pasteurisation Hot oil bath	Gamma-rays	Ultraviolet rays	Sieving through micro-filters	Acids Alcohol Formaldehyde Phenol Hypochlorite Mercuric chloride Ethylene oxide

cell components. Two commonly used methods of sterilisation by the use of heat are use of Hot Air Oven and Autoclave.

Hot Air Oven

It is a double walled insulated cabinet equipped with a heating element and fan for circulating hot air at a constant temperature. It also contains a temperature regulator and thermostat to adjust and maintain the temperature (Fig. 2.1). An oven is operated at a temperature of 160°C for one and a half to two hour for effective sterilisation. By increasing the temperature to 180°C, the time can be reduced to half an hour (Table 2.2). However, it is dangerous to sterilise glassware plugged with cotton or wrapped in paper at high temperature.

Table 2.2. Dry heat temperature-time relationship

Temperature (°C)	Time (min)
121	1080-1440
140	180
160	120
170	60
180	30

Oven is used to sterilise clean glassware such as Petri plates, pipettes, flasks and test tubes. Articles like Petri plates and pipettes are sterilised in metal cans made of copper, stainless steel or aluminium. The wide ends of pipettes are plugged with non-absorbent cotton wool before placing in long cans with plugged ends towards the mouth. The sterile pipettes are removed one by one under aseptic conditions.

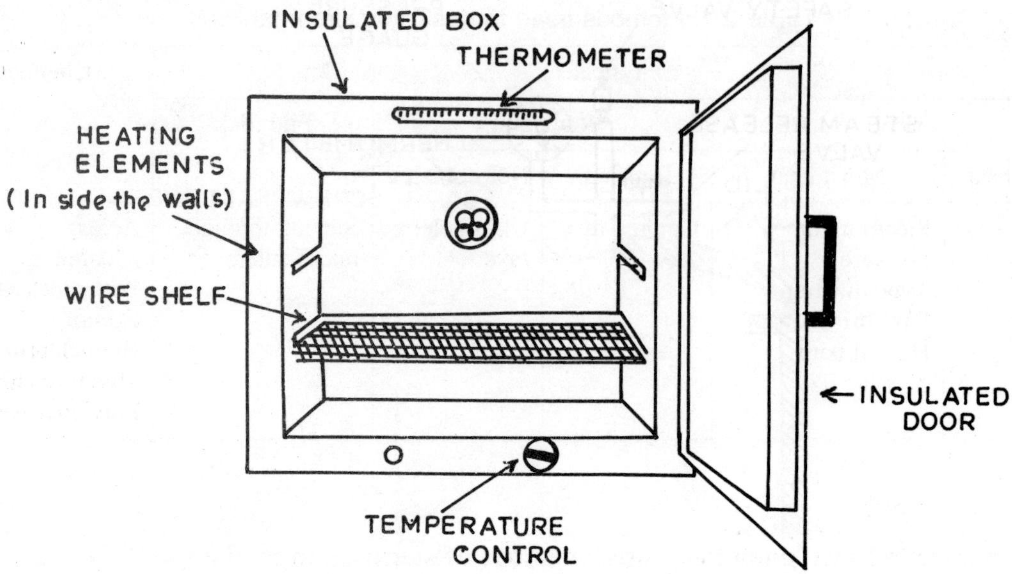

Fig. 2.1. The oven.

The Autoclave

The autoclave is a double walled cylindrical metal vessel, which can withstand a steam pressure of 50-60 lbs per square inch. There is a heating element in the base to generate steam from water. It has controls for adjustment of pressure, pressure gauze to measure the steam pressure and a thermometer to check the temperature inside the autoclave (Fig. 2.2). It is also equipped with a safety valve to prevent explosion because of excessive steam pressure and a steam outlet to evacuate air from the chamber. A stand or resting plate is also provided to keep the material to be sterilised.

Principle: The principle underlying sterilisation by autoclaving is that water boils at 100°C at atmospheric pressure. If the atmospheric pressure is increased, the boiling temperature of water will also increase. Thus if the steam pressure inside of the closed vessel is increased to 15 lb psi (2 atm), the temperature will rise to 121.6°C. All the organisms including heat resistant bacterial spore are killed if kept for 15-20 min at 121.6°C rendering all the treated material, sterile. Table 2.3 shows the relationship between pressure and temperature.

Operation: While operating, water is poured into the chamber so that the level of water is above the heating element and slightly below the resting plate. The articles to be sterilised are placed in the autoclave in a manner that steam can circulate freely between the articles placed inside. The lid of the autoclave is tightly screwed and heating element is switched on. As the water boils, steam is generated and rises to the top of the chamber forcing all the air out through the outlet valve. The outlet valve is closed when pure steam starts coming through it. The steam pressure starts rising which is indicated on the pressure gauze. At the required pressure of 15 lbs psi which gives a temperature of 121.6°C, the heating element is switched off automatically and further heating is controlled thermostatically. Check the temperature. The

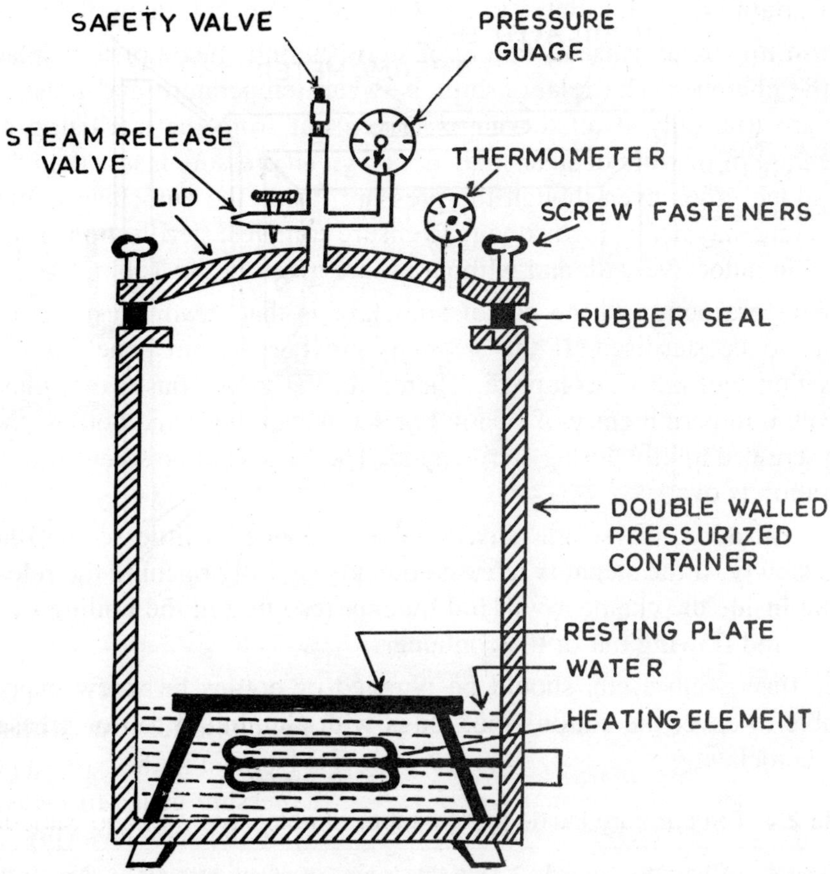

Fig. 2.2. The autoclave.

sterilisation begins only when the required temperature and pressure have been attained. The time of holding for sterilisation at 121.6°C is 15 minutes, however, the time varies with the size of vessel and amount of material it contains.

All liquids which are not heat sensitive and articles that are likely to be charred by dry heat can be sterilised in an autoclave.

Table 2.3. Relationship between pressure and temperature

Pressure (lbs psi)	Temperature (°C)
5	107.7
10	115.5
15	121.6
20	126.6
25	130.5
30	134.4

Precautions: Certain precautions have to be observed to prevent incomplete sterilisation.

1. The most important cause of failure of sterilisation is incomplete displacement of air from the chamber. The relationships between temperature and pressure mentioned above are true only if all the air is driven out from the sterilising chamber. The temperature of mixture of steam and air at a given pressure is less than that of the pure steam alone. Thus even though the pressure inside the autoclave is maintained, the temperature may not be high enough to ensure complete sterilisation. The temperatures attained in autoclave with and without air are given in the Table 2.4.
2. One more precaution in the use of autoclave is that steam must have access to the material to be sterilised. If the steam is prevented from penetrating the material, sterilisation will not be complete. Therefore, the tubes, flasks etc. must be plugged with cotton to permit entry of steam. For screw cap tubes and bottles, the tops should not be screwed tightly during sterilisation. The caps shall be screwed tightly when the sterilisation is over.
3. After sterilisation, the steam release valve is opened a little so that the steam may escape slowly. If the steam is released quickly by fully opening the release valve, the pressure inside the chamber will fall quickly resulting in the boiling of liquids in the chamber and flowing out of the containers.
4. All the flasks, tubes etc. should be plugged or bottles be screw capped. It is also advisable to cover the cotton wool plugs with aluminium foil or grease-proof paper before autoclaving.

Table 2.4. Temperature inside the chamber of autoclave with and without air

Gauze pressure (lbs)	All air discharged (°C)	Two-thirds air discharged (°C)	One-third air discharged (°C)	No air discharged (°C)
5	108	100	90	72
10	115	109	100	90
15	121	115	109	100
20	126	121	115	109
25	130	126	121	115
30	135	130	126	121

B. PASTEURISATION

Pasteurisation is not synonymous with sterilisation because all organisms are not killed in pasteurisation. The process was first devised by Pasteur to prevent the spoilage of wines. Today Pasteurisation is used to kill pathogenic bacteria responsible for causation of diseases like tuberculosis, brucillosis, Q fever and typhoid, frequently present in milk. Pasteurisation is performed by heating milk to 63-66°C for 30 min. or 71°C for 15 sec. The later one known, as Flash Pasteurisation is more prevalent in dairy operations since it does not alter the flavour of milk.

C. FILTER STERILISATION

Filtration involves the sieving of liquid or gas through a screen like material that has pores small enough to retain microorganisms. The microorganisms are retained partly because of small size of filter pores and partly by adsorption on the walls of pores while passing through the filter. Sterilisation by filtration has one drawback that it cannot retain viruses, since the size of virus particles is much smaller than the pore size of the filters (0.1-0.2 micron in diameter). During sterilisation by filtration, the screen or filter gets contaminated while the material that passes through it is sterilised.

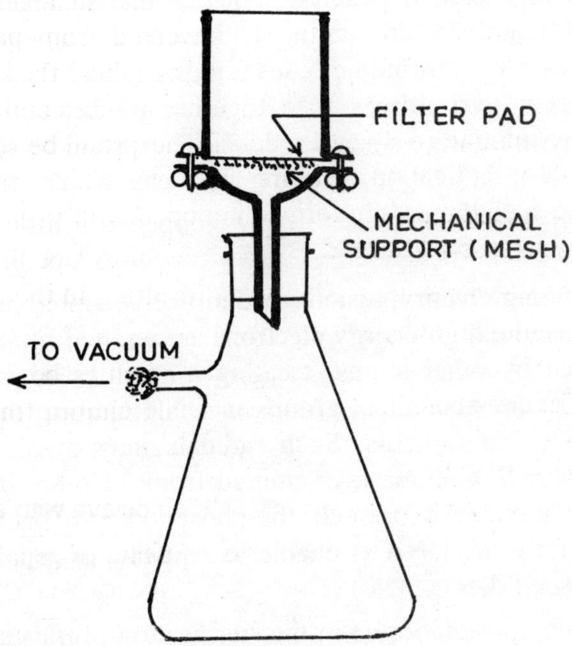

Fig. 2.3. Filter assembly.

The original filters were made of high quality ceramics, which were brittle and used to get clogged easily. Filters used to day are made of cellulose, asbestos or acetyl cellulose. Some commonly used filters are:

Seitz filters: Made of asbestos-cellulose mixture.

Sintered glass filters: Prepared by fusing fine glass fragments.

Membrane filters: Consists of porous discs of cellulose esters.

The filters are in the form of circular pads or discs, which are mounted on a special stainless steel funnel, and the mounting is fitted in a receiving flask. The whole assembly is sterilised in an autoclave before use. The solution to be filtered is poured onto the filter and either a positive pressure on unfiltered material or suction is applied to speed up the rate of filtration. The filtered material is collected and protected from contamination. When small amounts of solutions are to be sterilised, a membrane filter attached to a sterile syringe could be used.

Filtration is useful for sterilisation of substances that are heat sensitive. For example, heat sensitive substances such as vitamins, amino acids, enzyme solutions, cell extracts, etc. are sterilised only by filtration.

D. RADIATION STERILISATION

Ultraviolet Radiation: Microorganisms in the air are killed by exposure to ultraviolet (UV) radiation and thus has been a favourite method for sterilisation of large surface area or air space. UV radiation has poor penetration because it is strongly absorbed by media components. However, once reached to a cell, it penetrates the cytoplasm and is absorbed by purine, pyrimidine bases of DNA and aromatic amino acids such as tryptophan, phenylalanine and tyrosine etc. When absorbed by pyrimidine bases (cytosine and thymine) it causes changes in DNA, where two adjacent pyrimidines fuse together (dimerize) to give an intrastrand cyclobutane ring or a pyrimidine-(6-4)-pyrimidone. The presence of pyrimidine dimers in DNA strand prevents DNA replication resulting in a gap which must be repaired for the survival of the cell. If two such dimers are formed in opposite strands, the DNA molecule will break and cell may die.

Gamma Radiation: Ionizing electromagnetic radiation like alpha, beta, gamma and X-rays when absorbed by a molecule, high-energy electrons are emitted by its atoms. The electrons, thus ejected are absorbed by other atoms, creating a chain of ionization in the irradiated substance. This activity excites chemical groups in DNA causing the production of highly reactive, short-lived chemical radicals. Such radicals may break DNA strands causing mutations or death of the cell. Gamma rays emitted from ^{60}Co are of sufficient energy that when interact with DNA molecules in a cell, the phosphodiester backbone is broken. When enough breaks are produced, the DNA is unable to replicate or repair those phosphodiester bonds. Consequently the cell dies.

The major use of gamma irradiation is in the sterilisation of plastic wares like disposable Petri plates, tubes or hypodermic syringes. Gamma irradiation has also been used to prevent spoilage of packaged food without causing any significant changes in appearance, taste, texture or nutritive value of the product.

E. CHEMICAL STERILISATION

Another method that is useful for sterilisation of heat labile materials is the use of chemical agents. A chemical that kills or inhibits the growth of microorganisms is known as antimicrobial. The chemical agent that kills bacteria is called **Bactericidal** and the one that do not kill but inhibit the growth of bacteria is called **Bacteriostatic**. Likewise chemicals that kill or inhibit the growth of fungi are known as **Fungicidal** or **Fungistatic**. Antimicrobial agents generally known as **Germicides** are widely used where heat sterilisation is impractical. Such chemicals are used in hospitals for sterilisation of thermometers, lensed instruments, polyethylene tubing and catheters, inhalation and anesthesia equipment etc. For such operations cold sterilisation is performed in a close chamber that resemble autoclave but

Table 2.5. Items to be sterilised by various processes

Process	Item
Moist heat	Culture media, Water, Surgical instruments, Heat stable solutions, Rubber stoppers
Dry heat	Glass ware, Instruments, Needles, Oils
Ethylene oxide	Plastic tubing, Packaging materials, Implantable prosthetic devices, Polyethylene/plastic wares, Catheters, Diagnostic equipment
Ionizing radiation	Powders, Carriers; Bandages, Dialysis units, Surgical gowns
Filtration	Gases, Heat labile liquids, Low viscosity ointments, Oils

Table 2.6. Biological indicators used frequently

Spores of bacterial species	Sterilisation process	Approx. D-value (min.)
Bacillus subtilis	Ethylene oxide (600 mg l^{-1} at 50% RH and 54°C	3.0
Bacillus stearothermophilus	Moist Heat (≥ 121°C)	1.5
Clostridium sporogenes	Moist Heat (≥ 121°C)	0.2-0.8
Bacillus subtilis	Dry Heat (≥ 170°C)	0.8
Bacillus pumilus	Ionizing radiation	0.17 M rads

employ chemical agents such as ethylene oxide, formaldehyde or hydrogen peroxide. In food industry floors,walls and surfaces of equipment are treated with germicides to reduce microbial load. Drinking water is treated with chlorine to eliminate any potentially harmful organism.

In addition to their wide use in the medical field, disinfectants are widely used in industry to prevent microbial deterioration of food.

These chemical agents are highly reactive molecules that modify the molecules inside the cell resulting in destruction of their biological activities. Most commonly used agents are two gases **Ethylene oxide** or **Propylene oxide**. Both the chemicals boil at low temperature and are volatile. These chemicals act as alkylating agents and react with sulfhydryl and carboxyl groups of cell molecules. Ethylene oxide is in liquid form at and below 10°C. It boils at 10.7°C and turned into gaseous form. It is added to solutions in liquid form at concentrations varying from 0.5 to 1.0% at 0-4°C or is used in gaseous form at temperature above its boiling point. When used in gaseous form, the treatment is given in a special chamber in which temperature is below 50°C. Thus plastic materials can also be sterilised by treatment with ethylene oxide. Ethylene oxide is chemically unstable and decomposes to ethylene glycol that is non-volatile. Moreover, it is toxic to human being and is explosive. Thus it is used only in industries for sterilisation of heat labile materials.

Disinfectants and Antiseptics: Chemicals, which are used to kill microorganisms present on inanimate objects, are known as **Disinfectants** and chemical agents that kill or inhibit growth of microorganisms in living tissues are termed as **Antiseptics**. Such chemicals are non-toxic.

2.1. Physical Methods of Control: Heat

The bacteria differ in their tolerances to heat. When heat is applied, the bacteria are killed but only growth is inhibited at low temperature. Heat sensitivity is the intrinsic property of all organisms including bacteria and is determined genetically. It is partially reflected by the **optimum temperature** of growth. The bacteria are grouped into following categories on the basis of their optimal temperature.

Table 2.7. Categories of bacteria in terms of optimum temperature

Category	Minimum	Maximum	Optimum	Example
Psychrophilic	(–5) - (+5)	19-22	15-18	*Micrococcus ryophilus*, Cytophaga psychrophila
Mesophilic	10-15	35-47	30-45	*Escherichia coli*
Thermophilic	40-45	60-80	45-65	*Bacillus stearothermophius*
Hperthermophilic	65	95-105	88 or above	*Thermus aquaticus*

Bacteria are more heat resistant than most other forms of life. While non-photosynthetic bacteria can not only tolerate but can also multiply at and above 90°C, the upper limits of temperature tolerance for eukaryotic organisms range between 55-65°C. The optimum temperature for the bacterium *Pyridictium* is 105°C and that of *Thermococcus celer* is 88°C. Heat sensitivity of organisms can be affected by container size, cell density, moisture content, pH, and medium composition.

Heat can be applied as dry or moist heat. **Dry heat**, such as that in hot air ovens or incineration (e. g. flaming loops), denatures enzymes, dehydrates microbes, and killsorganisms by oxidation effects. **Moist heat** methods include pasteurisation, boiling, and autoclaving. In **Pasteurisation** the temperature is maintained at 63°C for 30 minutes or 72°C for 15 seconds to kill designated organisms that are pathogenic or cause spoilage. **Boiling** (100°C) for 10 minutes will kill vegetative bacterial cells, however, endospores are not inactivated. The most effective method of moist heat sterilisation is by **autoclave**, the use of steam under pressure. Increased pressure raises the boiling point of water and produces steam with a higher temperature. Standard conditions for autoclaving are 15 lbs. psi (121.6°C) for 15 minutes. This is usually sufficient to kill endospores and render materials sterile.

Effectiveness of heat for the control of microorganisms is measured by **Thermal death time** (TDP), the length of time required to kill all bacteria in a liquid culture at a given temperature and **Thermal death point** (TDP) the temperature required to kill all bacteria in a liquid culture in 10 minutes.

Objectives:

1. To compare the bactericidal effect of dry and moist heat.
2. To determine the heat tolerance of different microorganisms.

Materials required: Cultures of *Bacillus subtilis* (48-72 hrs Old culture), *Bacillus subtilis* (24 hrs Fresh culture), *Staphylococcus* spp., *Escherichia coli* and *Penicillium*, Petri plates containing nutrient agar, Thermometer, Empty tubes, Beaker, Hot plate, Ice and Autoclave.

Procedure:

1. Prepare three sets of broth culture of bacteria and spores of *Penicillium* spp. in separate sterile test tubes.
2. Prepare plates of nutrient agar and divide each plate into four sections with a marking pencil. Label each section as 0, 2.0, 5.0 and 15 min. Also write the name of organism and temperature treatment.
3. Set up two water baths using glass beakers. Dip two thermometers in water in separate test tubes and keep the test tubes into water baths. Note the temperature of water baths and maintain them at 63°C and 72°C.
4. Streak each organism on the surface of nutrient agar in respective plate and section marked as '0' time.
5. Place all the tubes containing different organisms in the water bath maintained at 63°C and 72°C.
6. Remove the test tubes after 2, 5, and 15 min incubation in the water bath and streak a loopful of each culture on the appropriate plate and section.
7. Autoclave one set of tube containing different organisms at 15 lb pressure for 15 min. and streak a loopful on the surface of nutrient agar.
8. Incubate all the plates in inverted position at 30°C for 48 hrs.

Observations and results: Examine the plates and record your results in the table as (0) = No growth, (+) = Minimum growth, (2+) = Moderate growth, (3+) = heavy growth, (4+) = Maximum growth.

Organism	Temperature °C/Time (min)											
	63°C				72°C				121.6°C			
	0	2	5	15	0	2	5	15	0	2	5	15
Bacillus subtilis (Old)												
Bacillus subtilis (Fresh)												
Staphylococcus spp.												
Escherichia coli												
Penicillium spp.												

2.2. Physical Methods of Control: Ultraviolet Radiation

Ultraviolet radiation though has poor penetrating ability, is used to sterilise some heat labile solutions, to decontaminate operation theaters, food processing areas and to disinfect wastewater.

Radiation of different wavelengths are emitted by the sun, some of which reach the earth planet Non-ionizing radiation between 15 and 390 nm is called **Ultraviolet (UV)**. Wavelengths below 200 nm are absorbed by air and do not reach the living organisms. The most lethal wavelengths are from 200-290 nm. These wavelengths are absorbed by DNA and cause damage. Ultraviolet light induces thymine dimers as a result of covalent bonding of two adjacent thymine molecules in one of the DNA strand in the nucleic acid and result in mutation. When mutation occur in critical genes, it results in the death of the cell. Pyrimidine dimers when exposed to visible light, photolyses are activated and dimers are split. This is known as **light repair or Photoreactivation**. Light repair is most effective in the wavelength varying from 420-540 nm. Another repair mechanism is **Dark repair** in which dimers are removed by endonuclease and DNA polymerase replaces the nucleotide bases. Finally the DNA ligase seals the exposed ends. Ultra violet radiation can not be used for sterilisation because of low penetration. It has practical application in surface sterilisation.

Objectives:

1. To examine the effect of ultraviolet radiation on bacteria.
2. To explain the method of action of ultraviolet radiation and light repair.

Materials required: *Escherichia coli*, *Saccharomyces cerevisiae*, Petri plates containing nutrient agar, Sterile cotton swabs, Ultraviolet lamp, Aluminium foil or Plastic sheet or Transparent glass plate.

Procedure:

1. Take a sterile swab, dip into one of the culture and spread on 6 plates of nutrient agar. To ensure complete coverage of the plate, swab the surface in two directions. Label the plates.
2. Remove the lids of the inoculated plates and cover one-half of each of the plate with one of the covering material.
3. Place one of the plate under the ultraviolet light approximately 24 cm. away from the light with lid removed and agar side up and expose in the following manner.
   (a) Expose two plates for 30 sec., replace the lid and incubate at 25°C.
   (b) Expose two plates more for 30 sec., replace the lid and incubate in the sunlight.
   (c) Expose two plates for 15 sec., replace the lid and incubate in dark in an incubator at 25°C.
   (d) Incubate all the plates at 25°C for 48 hrs.
4. Repeat the process with other organisms and compare the results.

Observations and results: Examine the plates and record your results in the table as (0) = No growth, (+) = Slight growth, (2+) = Moderate growth, (3+) = Abundant growth.

Organism	Time of Irradiation (min.)					
	0	1	3	5	7	10	15
Escherichia coli							
Saccharmyces cerevisiae							
Aspergillus niger							

2.3. Chemical Methods of Control: Disinfectants and Antiseptics

There is no single ideal chemical that can be used in all situations to kill or control the growth of microorganisms. Antimicrobial agent must match the specific organism and environmental conditions. Some of the factors that need to be considered while selecting antimicrobial chemicals are pH, solubility, toxicity and cost of the chemical. The criteria used to evaluate the effectiveness of antimicrobial agent are the concentration recommended by the manufacturer, length of contact needed and whether the chemical is lethal or inhibit the growth of microorganisms temporarily only.

Part A: Phenol Coefficient of a Disinfectant

The phenol coefficient test is a method to compare the antimicrobial activity of a chemical compound to that of phenol under standard conditions. A series of dilutions of the chemical to be tested and of pure phenol are prepared and placed into sterile test tubes in equal quantities. A standard quantity of a pure culture of the test microorganism, such as *Staphylococcus aureus* or *Salmonella typhi*, is added to each of the tubes. Subcultures of the test microorganisms are made from each dilution of the test chemical into sterile broth media at intervals of 5, 10, and 15 minutes after introduction of the organism. All the subcultures are incubated at 37°C for 48 hrs. and examined for the presence or absence of growth.

The phenol coefficient is determined by dividing the highest dilution of the chemical being tested that destroyed the microorganism in 10 minutes but not in 5 minutes, by the highest dilution of phenol that destroyed the microorganism in 10 minutes but not in 5 minutes. A phenol coefficient greater than 1 suggests that the chemical is more effective than phenol when employed under test conditions. A phenol coefficient of 5 indicates that the chemical agent under evaluation is five times as effective as phenol. Phenol coefficient is determined as follows:

Disinfectant		Growth (After incubation with disinfectant)		
		5 min.	10 min.	15 min.
Phenol Dilutions	1:50	–	–	–
	1:60	+	–	–
	1:70	+	+	–
Test Chemical Dilutions	1:250	–	–	–
	1:300	+	–	–
	1:350	+	+	–

Test chemical dilution of 1:300 showed no growth at 10 min. but grow at 5 min.
Phenol dilution of 1:60 showed no growth at 10 min. but grow at 5 min.

$$\text{Phenol coefficient} = \frac{300}{60} = 5$$

Objectives: To compare the effectiveness of disinfectants.

Materials required: Nutrient broth cultures of *Staphylococcus aureus* (24 hour), Nutrient broth tubes, Phenol, Lysol, Inoculating loop, Marking pencil and Bunsen burner.

Procedure:

1. Prepare dilutions of phenol (1:50, 1:60, 1:70) and lysol (dilutions: 1:250, 1:300, 1:350).
2. Label 18 nutrient broth tubes with the name and dilution of the disinfectant and the time interval of subculturing (e.g., phenol 1:50, 5 minutes).
3. In a test tube rack, place one test tube of each of the different phenol and Lysol dilutions.
4. Rapidly introduce one drop of the *S. aureus* culture into each of the test tubes of disinfectant. Note the time of introduction of microorganism into disinfectants.
5. Agitate all the test tubes to ensure contact of the disinfectant and the microbes.
6. Using sterile technique, at intervals of 5, 10, and 15 minutes transfer one loopful from each of the test tubes containing the disinfectant and microorganisms into the appropriately labeled sterile tube of nutrient broth.
7. Incubate all nutrient broth cultures for 48 hours at 37°C.

Observations and results: Observe the plates for growth, record the growth in the table and calculate phenol coefficient of each chemical.

Disinfectant		Growth (After incubation with disinfectant)		
		5 min.	10 min.	15 min.
Phenol Dilutions	1:50			
	1:60			
	1:70			
Test Chemical Dilutions	1:250			
	1:300			
	1:350			

Part B: Use Dilution Test

The standard method used for measuring the effectiveness of a chemical is **use-dilution test**, prescribed by American Official Analytical Chemists. The test organisms recommended are *Pseudomonas aerugenosa*, *Staphylococcus aureus* and *Salmonella choleraesuis*.

For use-dilution test dip metal rings into cultures of test bacteria grown in liquid media and allow to dry. The dried rings are then placed into solution of the disinfectant under test at the concentration recommended by the manufacturer after incubation for 10 min. at 20°C. Transfer the rings then to nutrient broth medium to permit the growth of surviving bacteria. The effectiveness of the disinfectant is determined by the amount of resulting growth. This test is limited to bactericidal compounds only and can not be used to evaluate bacteriostatic chemicals.

Objectives:

1. To evaluate the relative effectiveness of various chemicals as antimicrobial agents.
2. To describe the use-dilution test.

Materials required: Fresh culture of *Staphylococcus aureus*, Petri plates containing nutrient agar, Sterile Water, Sterile tubes, Sterile pipettes (1 & 5 ml) and Test chemicals such as Bathroom cleaner, Floor cleaner, Mouth wash etc.

Procedure: A simplified use-dilution test to be used in this exercise is described below.

1. Select a test compound and dilute in sterile water to a concentration recommended by the manufacturer. Add 5 ml in a sterile test tube.
2. Take two plates of nutrient agar and divide into five sections with a marker. Label these sections as 0, 2, 5, 10 and 20. Write down the name of disinfectant to be used also.
3. Inoculate a loopful of *S. aureus* in the section marked ‘0’.
4. Transfer 0.5-1 ml of actively growing culture of *S. aureus* in tube prepared in step 1.

5. Transfer one loopful from the tube to the corresponding section after 2, 5, 10 and 20 min.
6. Repeat all the above steps with other chemical compounds.
7. Incubate all the plates in inverted position at 35°C for 24-48 hrs.

Observations and results: Observe the plates for growth and record the growth in the table from 0 to +4 scale. (–) = No growth, (+) = minimum growth, (2+) = Moderate growth, (3+) =Heavy growth, (4+) = Maximum growth.

Time of Exposure (min)	Growth			
	Control	Disinfectant A	Disinfectant B	Disinfectant C
0		–	–	–
2	–			
5	–			
10	–			
20	–			

2.4. Chemical Methods of Control: Antimicrobial Drugs

In 1928 Alexander Fleming found that culture filtrates of *Penicillium* inhibited the growth of many Gram positive cocci and *Neisseria* spp. Later Selman A. Waksman isolated the antibiotic streptomycin, produced by an actinomycete. This antibiotic was found to be inhibitory to many bacteria that were not inhibited by penicillin. Even today, actinomycetes are important source of antibiotics and research workers look for antibiotic producing new strains of actinomycetes and fungi in soil. Many antimicrobial substances have also been synthesised in the laboratory. All antimicrobial chemicals absorbed or used internally, whether natural (antibiotics) or synthetics are called **chemotherapeutic agents**.

In order to treat an infectious disease, the concentration of chemotherapeutic agents has to be determined. For this the pathogen from a clinical sample is isolated and its sensitivity to chemotherapeutic agent is determined by **disk-diffusion method**. In this method, a Petri plate containing an agar growth medium is inoculated with the test organism uniformly over its entire surface. Paper disks impregnated with various chemotherapeutic agents are placed on the surface of the agar. During incubation, the chemotherapeutic agent diffuses from the disk (from an area of high concentration to an area of lower concentration). An effective agent will inhibit bacterial growth, and zone of inhibition around the disk is measured. The concentration of chemotherapeutic agent at the edge of the zone of inhibition represents its **minimum inhibitory concentration** (MIC). The MIC is determined by comparing the zone of inhibition with MIC values in a standard table (Table 2.8). The zone size is affected by such factors as the diffusion rate of the antibiotic and the growth rate of the organism. In this test *Mueller-Hilton agar* is used that allows the chemotherapeutic agent to diffuse freely.

Objectives:

1. To perform antibiotic sensitivity test.
2. To determine the chemotherapeutic effect of some chemicals and determine their minimum inhibitory concentration (MIC).

Materials required: Test organism, Petri plates containing *Mueller-Hilton* agar, Sterile cotton swabs, Antimicrobial disks, Forceps, Alcohol, and Ruler.

Procedure:

1. Dip the cotton swab into broth culture of test organism and inoculate onto the appropriate plate. Swab uniformly in three directions to ensure complete coverage of the agar medium. Let it stand for at least 5 minutes.
2. Sterilise forceps by dipping in alcohol and burning off the alcohol. Pick a disk impregnated with a chemotherapeutic agent with sterile forceps and place it on the surface of the agar approximately 2 cm from the edge of the plate. Gently tap the disk with the forceps to ensure better contact with the agar.
3. Repeat placing five to six different disks impregnated with different concentrations of chemical compounds to be tested on each Petri plate. Keep the disks at sufficient distance apart on the agar medium.
4. Record the agents and the disk concentration.
5. Incubate the plates in inverted position at 35°C for 24-48 hrs.

Observations and results: Observe all the plates for the presence of zone of inhibition surrounding each disk and measure the zones of inhibition in millimeters, using a ruler on the underside of the plate and record the observations in the chart as absence (0) or presence of zone of inhibition (+).

Based on the values in table determine whether the organism is sensitive, intermediate, or resistant.

Bacterial spp.	Antimicrobial Agent			
	A	B	C	D
Escherichia coli (Gram –ve)				
Staphylococcus aureus (Gram +ve)				
Bacillus cereus (Spore former)				
Mycobacterium smegmatis (Acid fast)				

Table 2.8. Interpretation of inhibition zones of test culture

Chemotherapeutic Agent	Disk Content (µg)	Diameter of Zones of Inhibition (mm)		
		Resistant	Intermediate	Susceptible
Ampicillin when testing G –ve bacteria (AM)	10	<13	14-16	>17
Ampicillin when testing G +ve bacteria (AM)	10	<28	–	29
Aztreonam (ATM)	30	<15	16-21	>22
Bacitracin (B)	10 units	<8	9-12	>13
Chloramphenicol (C)	30	<12	13-17	>18
Ciprofloxacin (CIP)	5	<15	16-20	>21
Erythromycin (E)	15	<13	14-22	>23
Sulfisoxazole (Gentrisin, G)	300	<12	13-16	>17
Imipenem (IPM)	10	<13	14-15	>16
Penicillin G when testing Staphylococci (P)	10 units	<28	–	>29
Penicillin G when testing other bacteria (P)	10 units	<14	–	>15
Satreptomycin (S)	10	<11	12-14	>15
Trimethoprim/Sulfamethoxazole (SxT)	1.25/23.75	<10	11-15	>16
Tetracycline (TE)	30	<14	15-18	>19
Vancomycin (VA)	30	<9	10-11	>12
Vancomycin when testing enterococci (VA)	30	<14	15-16	>17

From National Committee for Clinical Laboratory Standards

REVIEW QUESTIONS

1. Differentiate the terms bacteristatic and bactericidal.
2. Endospores of *Bacillus* spp. are more resistant as compared to the spores of fungi, why?
3. Discuss the detrimental effects of control agents on cell wall, nucleic acids and cytoplasmic membrane.
4. Compare the effect of hypertonic, hypotonic and isotonic solutions on microbial cells.
5. How hypertonicity is useful in the control of microbial growth?
6. How do halophiles survive in their environment?
7. Explain why ultraviolet rays cannot be used for sterilisation?
8. Explain why x-rays can be used for sterilisation and ultraviolet rays can only be used for surface disinfection?
9. Antibiotics like tetracycline, streptomycin and chloramphenicol inhibit the growth of bacteria but do not affect eucaryotic cells such as fungi and protozoa, why?
10. What do you understand by Phenol coefficient?
11. What is the effect of temperature on enzymes?
12. Where in nature would you most likely find a thermophile?

Chapter 3

Microscopy

Antonie van Leewenhoek who discovered microorganisms and laid the foundation of microbiology used a **Simple microscope** and described bacteria and protozoa for the first time in 1674. The microscope in use today is a **Compound microscope**. It contains a series of lenses and thus is much more powerful than the simple microscope. Further inclusion of Abbe condenser and oil immersion lenses led to the refinement of the compound microscope.

TYPES OF MICROSCOPES

Depending upon the principle on which the magnification of a microscope is attained, microscopes are of two types i.e. **Light microscope** and **Electron microscope**. While in light microscope the magnification is obtained with the help of light and optical lenses system, a beam of electrons is used in place of light to produce a magnified image in electron microscope.

A. Light Microscope

- Bright field microscope
- Dark field microscope
- Phase contrast microscope
- Ultraviolet microscope
- Fluorescence microscope

Bright Field Microscope

In routine examination of microorganisms, the most widely used microscope is bright field microscope. In this type of microscope, light rays pass directly through to the eye without being deflected by any intervening object. Here the microscopic field or area is brightly lighted and the objects being studied appear darker. This type of microscope produces magnification of about 1000 times which further can be increased by 2-3 times with little modification.

Dark Field Microscope

In dark field microscope, the microscope is slightly modified in such a manner that contrast of the image is increased against its background as a result of image being formed from diffracted, reflected and refracted light (Fig. 3.1). The principle behind this microscope is based on the observations of Tyndall that suspended particles in the path of a beam of light scatter the light at sharp angles to the original beam. As a result, a beam of sunlight highlights the particles in the air that otherwise are not visible. Similarly, if a microscope is fitted with a special condenser to block rays of light that normally hit the objective lens, the observer see a dark field. If, however, any cell is present, it will scatter light and scattered rays of light will produce an approximate image of the cell. The dark field procedure is performed by fitting a dark field stop below Abbe condenser. This eliminates light from the center and a thin cone of light rays reaches the specimen at an angle to the objective lens.

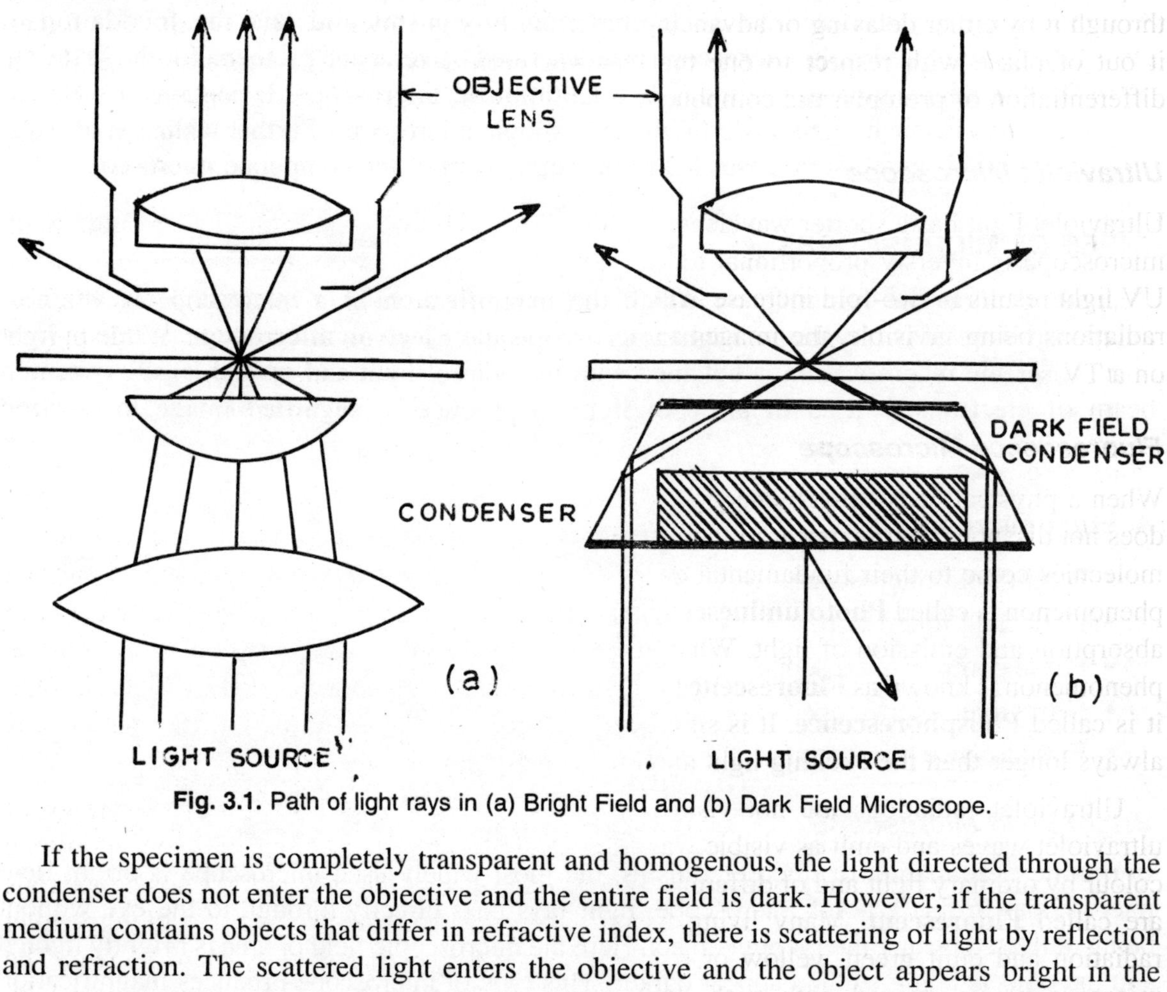

Fig. 3.1. Path of light rays in (a) Bright Field and (b) Dark Field Microscope.

If the specimen is completely transparent and homogenous, the light directed through the condenser does not enter the objective and the entire field is dark. However, if the transparent medium contains objects that differ in refractive index, there is scattering of light by reflection and refraction. The scattered light enters the objective and the object appears bright in the otherwise dark field. Dark field microscopy is valuable in the examination of unstained microorganisms suspended in fluid as wet mount and hanging drop preparations.

Phase Contrast Microscope

A phase contrast microscope is capable of differentiating transparent protoplasmic structures that differ only slightly in their refractive indices. The phase contrast microscope was discovered by Fritz Zernike in 1933 and was awarded Nobel prize in physics in 1953 for this invention.

Here an annular ring present below the condenser allows only a ring of light to pass through the condenser and strike the specimen. The microscopic objective also contains a transparent disc called phase-shifting plate. A ring on this disc is used to focus the light from annular diaphragm. Depending upon its composition, this ring can alter the phase of light wave passing through it by either delaying or advancing them. By briefly slowing one wave the disc forces it out of phase with respect to one that has not been delayed. This phase shift results in differentiation of protoplasmic components of slightly different refractive indices.

Ultraviolet Microscope

Ultraviolet light has a shorter wavelength (180-400 nm) than visible light. Since resolution of microscope is inversly proportional to wavelength of light used for illumination, the use of UV light results in two-fold increase in resolution over bright field microscope. However, UV radiations being invisible, the images are recorded on a photographic plate or are displayed on a TV screen.

Fluorescence Microscope

When a physical body absorb energy, its molecules are activated. The energy thus absorbed does not disappear but reappear in the form of chemical reaction, heat or light and the excited molecules come to their fundamental energy state. If the energized molecules emit light, the phenomenon is called **Photoluminescence**. In photoluminescence, there is time lapse between absorption and emission of light. When the time lapse is less than 1/10000 of a second, the phenomenon is known as **Fluorescence** and when time lapse is more than 1/10000 of a second, it is called **Phosphorescence**. It is significant to note that the wavelength of fluorescence is always longer than the exciting light and thus is generally visible.

Ultraviolet radiations are not visible. However, some chemicals absorb the energy of ultraviolet waves and emit as visible waves of greater length. Thus a material appears of one colour by ordinary light and of different colour in presence of ultraviolet light. Such materials are called **Fluorescent**. Many living tissues fluoresce naturally in presence of ultraviolet radiation and emit green, yellow or red light. Several dyes that fluoresce are known. The microorganisms stained with such dyes appear as luminous objects when observed by ultraviolet illumination. One such dye is **Auramine-D**, which glows yellow in presence of UV light. It has selective affinity for wax like substances present in *Mycobacterium tuberculosis*. **Immunofluorescence** is another application of this type of microscopy. In this, fluorescent dyes are chemically attached with blood serum protein (antibodies). When such antibodies react with antigen (bacterium) against which they have been raised will give fluorescence in UV light and can be detected by microscopic examination. Two such dyes known are **Fluorescein isothiocyanate** and **Lissamine rhodamine B** which give apple green and orange colour, respectively.

B. Electron Microscope

The fact that light microscopes with resolution less than 0.2 μ could not be prepared, led to the discovery of electron microscope by Ernst Ruska and Max Knoll in early thirties.

Electrons are very small negatively charged particles and behave like light waves. Electrons when accelerated to 100000 volts produce a wavelength of 0.05 Å, which is much shorter than even the UV light. Thus these electrons when used to illuminate an object can provide a resolution of 4 Å or 0.0004 μ. Thus the lens and illumination system in electron microscope are similar to those of the light microscope except that in electron microscope the specimen is illuminated by a beam of electron rather than light and the focusing is carried out by electromagnets instead of a set of optical lenses. All these components are sealed in a tube in which a complete vacuum is established. An electron microscope can provide a magnification upto one million time that permit visualisation of submicroscopic cellular particles and viral structures. However,electron beams do not penetrate the cell well and to examine the internal cell structure, thin sections are needed. Thus special techniques are employed to prepare specimen for electron microscope. To obtain sufficient contrast, special electron stains such as osmic acid, permanganate, uranium, lanthanum or lead are used. These substances are composed of atoms of high atomic weight and scatter electrons well, thus improve the contrast.

THEORATICAL PRINCIPLES OF MICROSCOPY

To use the microscope with efficiency, one must understand the basic principles of microscopy like Magnification, Resolution, Numerical Aperture, Illumination and Contrast etc.

Magnification

The ratio of image to object is called magnification. In a compound microscope magnification is obtained by a series of two lens systems, the objective and ocular or eye piece. Objective magnifies the specimen and produces a real image, which is further magnified by ocular yielding a virtual image. The eye sees this virtual image. The total magnification is equal to the product of the magnification of ocular and the magnification of objective. The initial magnifications of objective and ocular are engraved on the objectives and oculars.

	Magnification		Total magnification
	Objective	Ocular	Objective × Ocular
Scanning	5×	10×	50×
Low power	10×	10×	100×
High and dry	45×	10×	450×
Oil immersion	100×	10×	1000×

Resolution

Magnification is important but it is not possible to increase the magnification merely by increasing the magnifying power of the lenses or by use of additional lenses. Because lenses are limited by a property called **Resolution**—the ability of the lens to distinguish two adjacent points as distinct and separate. It means that the larger magnification produced by a microscope may not be useful because image obtained may be unclear or fuzzy. The resolution of a lens is dependent on the wavelength of light used and another characteristic of the lens called **Numerical aparture.**

$$\text{Resolution} = \frac{\text{Wavelength}}{\text{Numerical aperture}}$$

Accordingly shorter the wavelength, greater the resolution of the lens. Thus light rays of short wavelengths are better suited than that of longer wavelengths. However, merely by decreasing the wavelength we cannot increase resolution because visible portion of the electromagnetic spectrum is narrow (400-700 nm).

Numerical aperture

The angle (θ) subtended by the optical axis and the outermost rays still covered by the objective lens is a measure of the aperture of the objective. It is one half the aperture angle (Fig. 3.2).

$$\text{Numerical aparture (NA)} = n \sin \theta$$

where n = referective index of medium.

When dry objective is used, the value of n is 1, since 1 is the referective index of air. When oil immersion lens is used, the value of n is 1.56 because referective index of cidar wood oil is 1.56. If θ is 60°

$$\begin{aligned} \text{NA} &= n \sin \theta \\ &= 1.56 \times \sin 60 \\ &= 1.56 \times 0.9 \\ &= 1.40 \end{aligned}$$

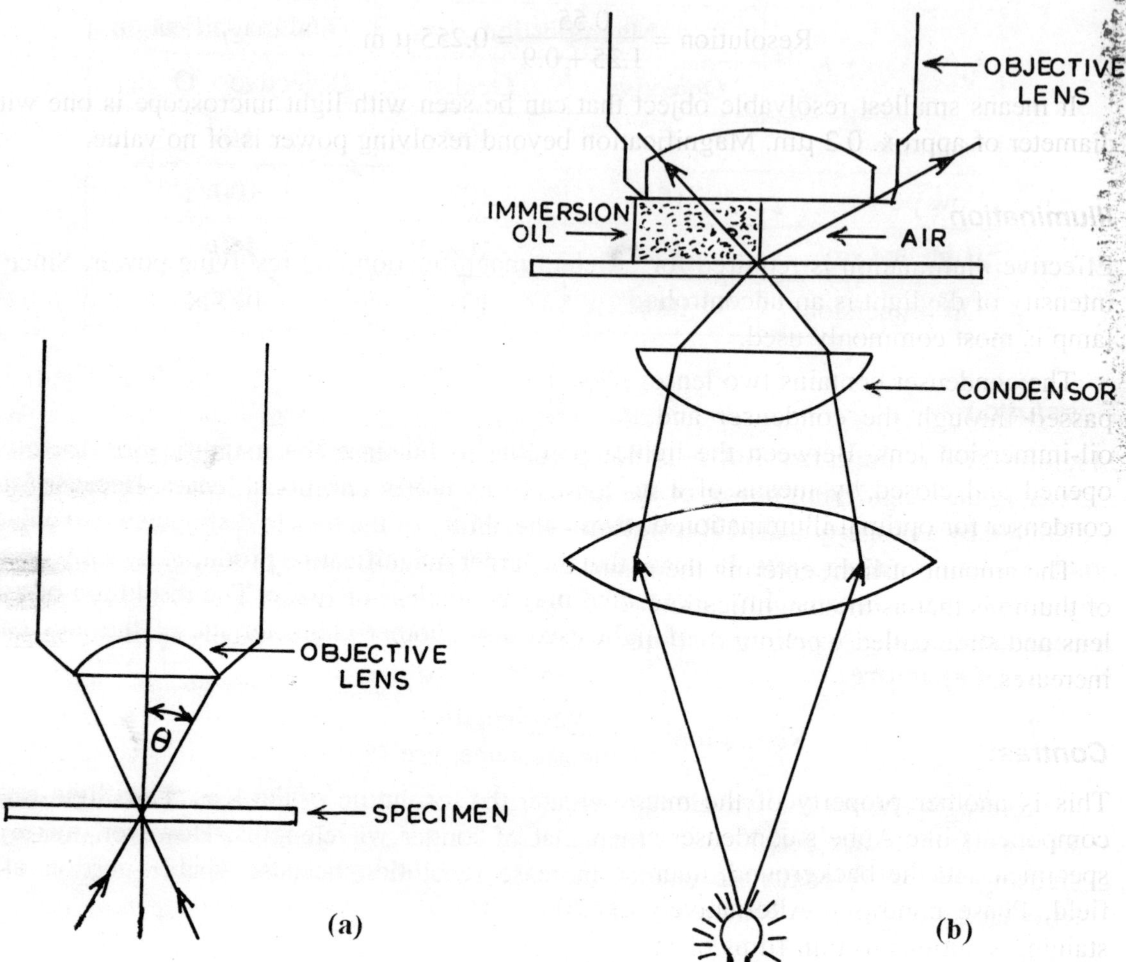

Fig. 3.2. (a) Angle subtended by incident rays entering the objective. (b) Path of light rays with and without immersion oil.

Microscope objective can be altered to increase the NA to limited extent. The maximum NA for a dry objective is less than 1.0 and for oil-immersion objective its range is 1.2-1.4. Since wavelength of visible light range between 400 and 700 nm the resolving power of a microscope is restricted to the limited values of NA and wavelength of visible light.

Numerical aperture of an objective can be increased by the use of substage condenser that illuminates the object with rays of light. However, the maximum limit is 2 fold.

$$\text{Resolution} = \frac{\text{Wavelength of light } (\lambda)}{\text{NA (objective)} + \text{NA (condensor)}}$$

If $\lambda = 0.55\ \mu\text{ m}$

NA (Oil immersion) = 1.25

NA (Condensor) = 0.9

$$\text{Resolution} = \frac{0.55}{1.25 + 0.9} = 0.255\ \mu\text{ m}$$

It means smallest resolvable object that can be seen with light microscope is one with a diameter of approx. 0.2 μm. Magnification beyond resolving power is of no value.

Illumination

Effective illumination is required for efficient magnification and resolving power. Since the intensity of daylight is an uncontrolled variable, an artificial source of light such as tungsten lamp is most commonly used.

The condenser contains two lenses to produce maximum numerical aperture. The light is passed through the condenser and it is kept close to the stage especially when using oil-immersion lens. Between the light source and condenser is Iris diaphragm that can be opened and closed by means of a lever, thus regulating the amount of light entering the condenser for optimal illumination of the object.

The amount of light entering the microscope differs with each objective lens used. A rule of thumb is that as the magnification of the lens increases, the distance between the objective lens and slide called **working distance**, decreases and the angle of aperture of the objective increases.

Contrast

This is another property of the microscope for effective visualation of specimens. While components like Abbe's condenser and Iris diaphragm help to increase contrast between the specimen and the background, microscopes have been modified to improve contrast (Dark field, Phase contrast). Alternatively cells and cellular structures are stained by applying staining solutions to thin films of microorganisms deposited on clean glass slide to increase contrast. These stained preparations are examined for the presence of internal and external structures.

Aberrations

The rays of light pass through objective lens, bend, converge at the focal point and an image is formed. When an object present in the center of microscopic field is focused, the object present in the periphery may not be in focus and the image looks fuzzy. This is because of curvature of the lens that forms multiple focal points. This is known as **Spherical aberration** and can be minimized by the use of Iris diaphragm that eliminates light rays to the periphery of lens. Another type of aberration encountered is **Chromatic aberration**. The lens system in the microscope acts like a prism and splits light into different wavelengths resulting in the formation of different focal points for each wavelength. Such aberrations are called chromatic aberrations and can be minimized by the use of filters. Usually a blue filter is recommended. Alternately a light source of one wavelength or monochromatic light can be used to eliminate chromatic aberrations. However, such light sources are expensive.

3.1. Bright Field Microscope and Its Use

Components of a Compound Microscope

The basic frame of the microscope consists of a **base**, an **arm**, a **stage** and a **body tube**. All other vital parts are attached to the frame of the microscope (Fig. 3.3). In older microscopes the base is not fixed to the arm instead is joined with a pivot which enables the user to tilt the arm backward to adjust the eye point height.

Stage

A horizontal platform with an opening in the center for passage of light from illumination source to the object is called stage. It provides a platform for the specimen. It has clamping device or mechanical stage for holding and moving the slide.

Source of Illumination

Some microscopes have built in source of light in the base while others have a **mirror** with one side flat and other concave. Most of the microscopes have a blue **filter** over the light source which provide short wavelength blue light for better resolution. The microscopes with built in electric source of light generally have provision for voltage control to vary the intensity of light.

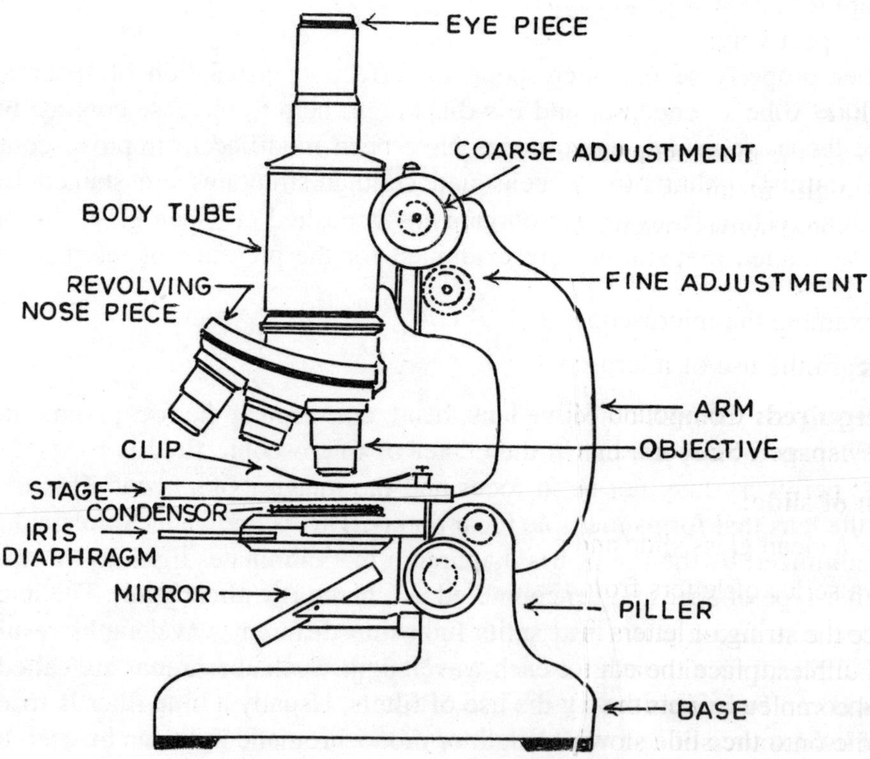

Fig. 3.3. The compound microscope.

The microscopes that have mirror are used by placing either near a source of light (flat surface) or near a window with natural sunlight (concave surface).

Abbe condenser

Below the stage are two sets of convex lenses to collect and concentrate light rays coming from light source for proper illumination of the object. Condenser is equipped with **Iris diaphragm**, which is a shutter controlled by a lever and regulates the amount of light entering the lens system.

Body tube

Above the stage and attached to the arm is body tube. This structure houses the lens system that helps to magnify the specimen.

Lens system

All microscopes have three lens systems—the oculars, the objectives and the condenser. The **Ocular** or **eyepiece** is attached to the upper end of the body tube. It consists of two or more lenses. Oculars of varying magnification (10× or 15×) are interchangeable. On the lower end of body tube is attached a revolving Nosepiece which can accommodate 3-4 **Objectives.** Objectives on most microscopes have magnification of 5×, 10×, 45× and 100×. Objectives of 5× and 10× are termed **low power** and that of 45× and 100× are known as **High dry** and **Oil immersion**, respectively.

Focusing knobs

The body tube of the microscope can be raised or lowered with the help of **Coarse adjustment** and **Fine adjustment** knobs for bringing the objects into focus while examining a slide. In some microscopes stage is lowered or raised in place of body tube for the purpose of focussing.

Objectives:

1. To examine the microscope and locate the various components.
2. To learn the use of microscope.

Materials required: Compound Microscope, Glass slides, Cover slips, Lens paper, Xylene, Dropper, Newspaper, Scissors, Inoculating needle

Preparation of slide:

1. Take a clean glass slide and cover slip and place a drop of water on the center of slide.
2. Cut a series of letters from the newspaper including small 'e'.
3. Place the string of letters into the drop of water and cover with the cover slip. To avoid air bubbles, place the edge of the cover slip on the drop of water on the slide so that fluid completely wets the edge. Then lower the cover slip with the help of a inoculation needle onto the slide slowly and slowly.
4. Examine and make a sketch of the letter 'e' as it appears on your slide.

5. Examine the preparation under the low power and high power objectives as per procedure for operation of microscope given below.

Operation of microscope:

1. Wipe all lenses with lens paper.
2. Move the low power objective into position under the body tube. You will feel a click when it is in place.
3. Connect the light source and turn it on. If the microscope does not have a built in light source but a mirror, place the microscope near a well-illuminated window.
4. Place the slide prepared on the microscope stage and secure it with clips. Arrange the portion of the preparation to be examined over the central opening in the stage.
5. Lower the low power objective with the help of coarse adjustment knob to approximately 1 cm above the cover slip of your preparation.
6. Raise the condenser as far as it will go with the aid of the condenser adjustment knob.
7. Look through the ocular. Move the iris diaphragm lever so that the diaphragm is open to its maximum limit. This is determined by looking through the eyepiece and noting the changes in the intensity of light.
8. Focus by movement of body tube with coarse and fine adjustment knobs.
9. The microscope stay in focus with all the objectives. Thus you need not raise the body tube to change the objectives (change from low power to high power objective). This property of microscope is known as par focal.

Care of the microscope: Some precautions need to be observed for the successful operation and proper maintenance of the microscope are:

1. Hold the arm of the microscope with one hand and place the other hand under the base to provide support while carrying the instrument. Keep the instrument in an upright position.
2. Remove the mirror before lifting the microscope.
3. Keep the microscope at least six inches away from the edge of laboratory table.
4. Do not handle the lenses with your fingers. Perspiration contains fatty acids and other substances, which damages the lenses.
5. Use lens paper for cleaning the lenses
6. Wipe the lenses of the microscope before and after use
7. Do not tamper with the components of the microscope. In case of any problem, report to the teacher immediately.
8. Do not apply chemicals to any part of the microscope.
9. Remove immersion oil from oil immersion objective with lens paper

10. Be sure that low power objective is in the working position before putting the microscope into cabinet
11. If a cover is provided for the microscope, be certain that it covers the entire instrument

3.2. Microscopic Measurements

Being minute in size, the dimensions of microorganisms are expressed in microns (micrometers). A micrometer is one thousandth of a millimeter and one millionth of a meter. The dimensions of viruses that are much smaller than bacteria are measured in nanometer and angstrons. A nanometer is one billionth of a meter and an angstron is one tenth of a nanometer.

Microorganisms other than viruses vary in size from a fraction of micrometer to multiple of a micrometer. The size of such microorganisms can be measured by equipping the microscope with **Ocular micrometer** and **Stage micrometer**.

Ocular micrometer: An ocular micrometer consists of a circular glass disk, which has graduations, engraved on its upper surface (Fig. 3.4). The graduations on the ocular micrometer are equally spaced and marked 0-100 although these graduations do not have any standard value. The distance between graduations varies depending upon the magnification of ocular used. The value of each division of ocular micrometer can be determined by calibrating it with a known scale of stage micrometer. Ocular micrometer is put inside the eyepiece by disassembling the ocular.

Stage micrometer: It is a precision made glass slide which contains one millimeter line equally divided into 100 divisions. Thus each division measures 0.01 mm (10 micron). The distances between each division of stage micrometer become correspondingly enlarged under low or high power or oil immersion objective. Stage micrometer is used for calibration of ocular by by super imposing the graduations of one over the other.

Calibration of Ocular Micrometer

For microscopic measurements, ocular micrometer scale has to be calibrated with a standard scale of stage micrometer for each type of objective. Microscopes also differ in their body tube length slightly. Thus the magnifying power of each microscope differ though slightly. Therefore, it is essential to standardise ocular micrometer, if the microscope is changed.

Objectives:

1. To get familiar with ocular and stage micrometers,
2. To calibrate ocular micrometer using stage micrometer,
3. To measure the size of microbial cell.

Materials required: Ocular micrometer, Stage micrometer, Microscope, Clean slides, *Saccharomyces cerevisiae* and *Bacillus subtilis* culture broth.

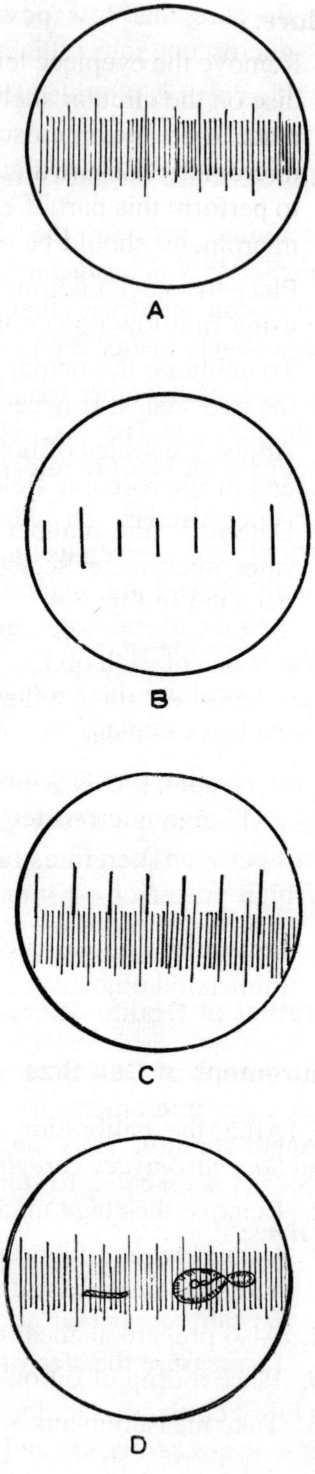

Ocular Micrometer
Graduations are arbitrary but equidistant.

Stage Micrometer
Graduations are spaced 0.01 mm (10 m) apart.

Calibration of ocular micrometer
Ocular micrometer superimposed over stage micrometer.

Measurment of microorganisms
Calibrated ocular micrometer is used after removing stage micrometer.

Fig. 3.4. Stage and ocular micrometers.

Procedure:

1. Remove the eyepiece lens from the microscope, unscrew and insert ocular micrometer disc on the circular shelf inside the eye piece. Tighten the screw of eyepiece back and place it in the microscope. When observed, you can see the scale line of ocular micrometer. Ensure that eye piece is not dropped or reassembled incorrectly. It is better to perform this part of exercise in presence of the instructor. Graduations of the ocular micrometer should be on the upper surface.
2. Place the stage micrometer on the microscope stage and focus the scale of micrometer using first low power and then high dry objectives.
3. To calibrate the ocular scale for any given objective, it is necessary to super impose the two scales. If necessary the ocular may be rotated slightly.
4. Adjust the scales of both micrometers in such a way that lines of two coincide at one end of microscopic field.
5. Determine the number of ocular divisions that coincide with one graduation of the stage micrometer scale. How the two scales appear when they are aligned properly is depicted in Fig. 3.4.
6. Repeat the steps 3 to 5 and take at least six measurements.

   Now taking the average of the six measurements or using three concordant readings we can calculate the measurement of each ocular division.

   Let 10 stage micrometer division = 18 ocular micrometer divisions
   1 stage micrometer division (known) = 10 micron
   Therefore, 18 ocular divisions = 100 microns
   1 ocular micrometer division = 100/18 = 5.56 microns

7. Repeat the above procedure and calibrate ocular micrometer using high and oil immersion objectives.

Measurement of Cell Size

1. After the calibration of ocular micrometer is complete, prepare a wet mount of *Saccharomyces cerevisiae* culture.
2. Remove the stage micrometer and place your slide on the platform of stage.
3. Using low and high-dry objectives measure the diameter of a circular cell or length and diameter of a rod shaped cell in terms of ocular divisions.
4. Also prepare a smear of *Bacillus subtilis* and stain using simple staining technique.
5. Place a drop of oil on the smear and focus using oil immersion objective.
6. Take measurements and record your observations.

Observations:

Microscope Make and No.	
Objective	Value of each ocular division (μ)
Low power	
High-dry	
Oil immersion	

Organisms	Objective lens	No. of ocular divisions		Actual dimensions (μ)	
		Length	Breadth	Length	Breadth
Saccharomyces cerevisiae	Low power				
	High dry				
Bacillus subtilis	High dry				
	Oil immersion				

3.3. Direct Microscopic Count of Microorganisms - The Breed Count

The number and type of microorganisms present directly affect the quality of water, food, milk or air. The total number of microorganisms in these substances can be measured by either direct or indirect counts. Though standard plate count is the most reliable method to determine the quality of water, food, milk or air, the direct microscopic count using a slide is performed when it is essential to count the number of microorganisms in any given sample quickly. This is accomplished by spreading a known amount of sample material on an area of one square centimeter of the slide—**Breed slide**. The slide is stained and examined under oil immersion objective and all the organisms in microscopic field are counted. For accuracy; the numbers of organisms are counted in several microscopic fields and an average number is determined for use in the calculations.

Breed Count Method

The Breed Count named after its originator though gives a rough estimate of microorganisms but is a useful technique to determine the microbiological quality of any given sample. Any suitable staining technique may be used i.e. either simple or Gram's staining. In case of milk, defatting is required. Thus Newman's stain is used which fulfill the purpose of both defatting as well as staining. However, in this method it is essential that the microscope is first calibrated to determine the area of microscopic field.

Materials required: Breed slide, Breed pipette (micro-pipette, 0.01 ml), Methylene blue, Stage micrometer, Microscope.

Calibration of microscope: Microscopic factor (MF) reveals what part of a milliliter of a sample each microscopic field represents. To determine MF, the area of microscopic field is calculated by determining the diameter of microscopic field with the help of a stage micrometer.

1. Set the microscope using 10× eyepiece and low power objective. Adjust and focus the stage micrometer.
2. Place a drop of cidar wood/paraffin oil on the stage micrometer and focus using oil immersion objective. Keeping in mind that each division of stage micrometer is equivalent to 0.01 mm, measure the diameter of the field.
3. Calculate the area of field in square millimeters for the oil immersion objective and 10× eyepiece using formula πr^2 ($\pi = 3.14$, r = radius).
4. Determine the microscopic factor as follows:

$$\text{MF} = \text{Number of fields in 1 cm}^2 \text{ (100 mm}^2\text{)}$$

$$= 100/\text{Area of field in mm}^2$$

$$n = \text{Average number of organisms or cells per field} = \frac{\text{No. of cells counted}}{\text{No. of fields counted}}$$

$$N = \text{Number of organisms or cells present in 1 cm}^2 = \text{MF} \times n$$

Total number of cells or organisms present in 1 ml of sample (since 0.01 ml of sample was spread over 1 cm^2 multiply by 100) = MF × n × 100

In general there are 5000 microscopic fields in one square centimeter and each field represents 1/50000 of a ml of sample.

3.4. Direct Microscopic Count of Microorganisms: Use of Neubauer Counting Chamber

Another quick and precise method to determine the number of microorganisms in any sample is the use of Neubauer counting chamber (Haemocytometer). The method is quick but needs expertisation and gives total (both viable and dead) number of cells.

Neubauer Counting Chamber: It is precision-machined glass plate that has a sunken platform at its center. The depth of this platform is 0.1 mm below the surface of the plate. The surface of the plate is etched with a grid system measuring 1 mm × 1 mm and consisting of 25 large squares marked with bold/triple lines. Each of the large/square is divided into 16 small squares. Thus an area of 1 mm^2 is divided into 400 small squares (Fig. 3.5). The total grid area when covered with a cover slip thus has a volume of 1 mm × 1 mm × 0.1 mm or 10^{-4} cm^3. The volume of each small or large square can also be calculated from the dimensions known. A more precise counting chamber available is **Petroff Hausser Counting Chamber**.

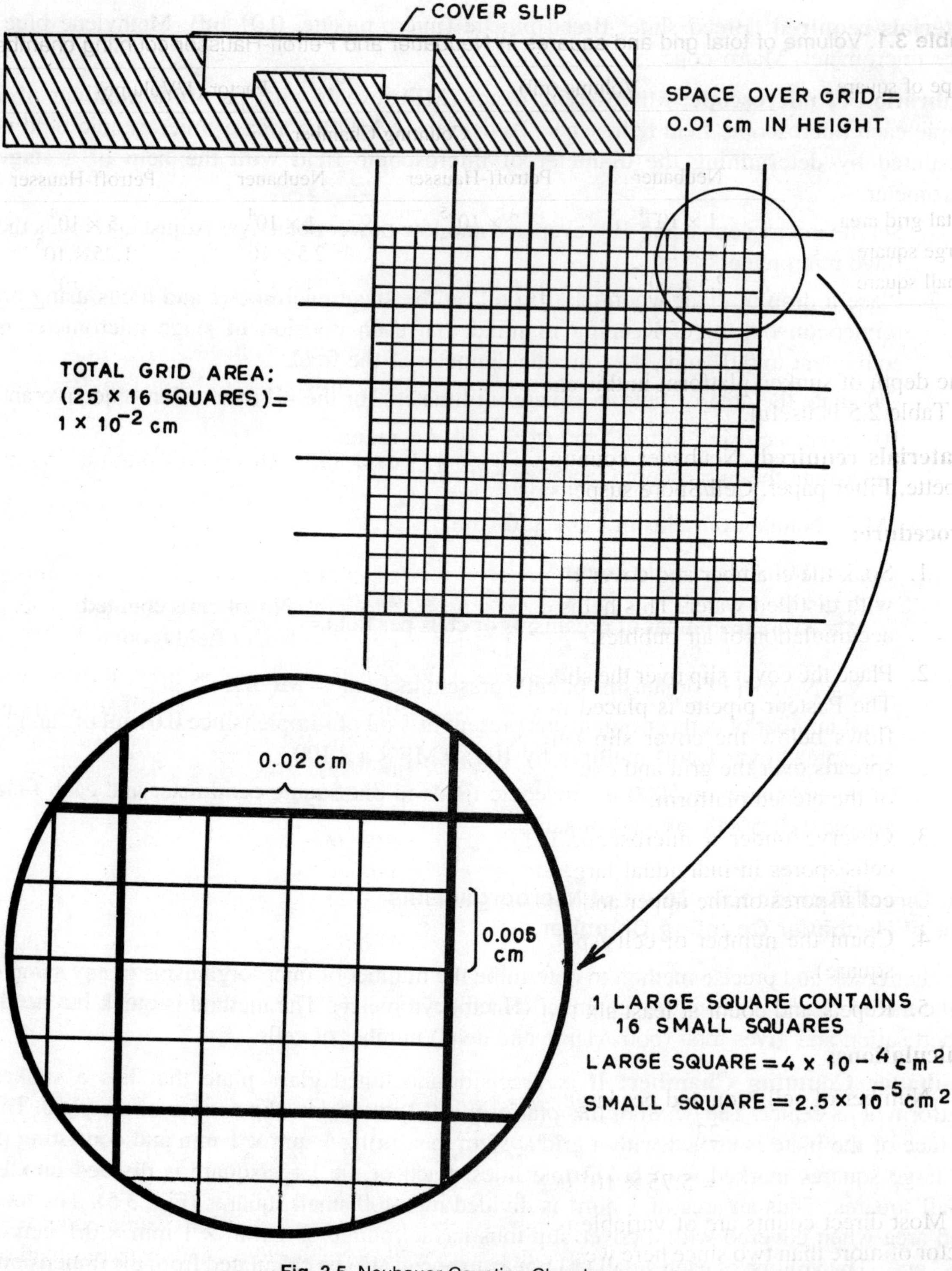

Fig. 3.5. Neubauer Counting Chamber.

Table 3.1. Volume of total grid and squares in Neubauer and Petroff-Hausser counting chamber

Type of square	Volume (ml)		Factor (1/Volume)	
	Counting Chamber			
	Neubauer	Petroff-Hausser	Neubauer	Petroff-Hausser
Total grid area	1×10^{-4}	2×10^{-5}	1×10^{4}	5×10^{4}
Large square	4×10^{-6}	8×10^{-7}	2.5×10^{5}	1.25×10^{5}
Small square	2.5×10^{-7}	5×10^{-8}	4×10^{6}	2×10^{7}

The depth of sunken platform in this chamber is 0.02 mm instead of 0.1 mm. The data given in Table 2.5 is useful.

Materials required: Neubauer counting chamber, Cover slips, Detergent solution, Pasteur pipette, Filter paper, Cell/Spore suspension

Procedure:

1. Soak the chamber and cover slip in a solution of mild detergent and rinse thoroughly with distilled water. This helps even flow of the liquid into the chamber and prevent accumulation of air bubbles.
2. Place the cover slip over the slide and introduce the liquid sample with Pasteur pipette. The Pasteur pipette is placed in contact with the slide in such a way that the liquid flows below the cover slip automatically by capillary action. The cell suspension spreads over the grid and excess culture overflows into the two channels at the edges of the etched platform.
3. Observe under a microscope using 10× or 40× objective. Count the number of cells/spores in individual large square. To avoid counting of same cells twice, omit cells/spores on the upper and left hand borderlines of each square.
4. Count the number of cells/spores in at least 10 large squares (10-20 cells per large square).
5. Repeat and count at least six times to obtain minimal variation.

Calculations:

Number of cells counted per large square × Dilution factor × Factor for large square

$$23 \times 10^{2} \times 2.5 \times 10^{5}$$

$$= 5.75 \times 10^{8} \text{ cells per ml}$$

Most direct counts are of variable reliability and they over estimate the viable-count by a factor of more than two since here we count dead as well as viable cells. Moreover, this method is suitable only for counting the cells in liquid culture and not in soil or any other particulate material.

REVIEW QUESTION

1. What is the total magnification obtainable with low power and high power objectives?
2. What happens to the original orientation of the letter 'e' when the slide is moved a) side to side, (b) up and down?
3. Distinguish between resolving power and par focal.
4. Write down the various components of a microscope and explain function(s) of each.
5. What controls the amount of light reaching the ocular lens?
6. What is meant by par focal in microscopy?
7. Name two ways in which the resolving power can be increased?
8. Explain the functions of each of the following in microscopy.
   (a) Iris diaphragm, (b) Coarse adjustment, (c) Fine adjustment, (d) Condenser
9. What do you expect if water drop is used in place of immersion oil?
10. What are chromatic aberrations?
11. Write the functions of Iris diaphragm and condenser.
12. Define numerical aperture.
13. How electron microscope resemble light microscope.
14. Differentiate dark field microscope from bright field microscope.
15. What for ocular and stage micrometers are used?
16. How Neubauer counting chamber differs from Petroff-Hausser counting chamber?
17. Direct microscopic count generally overestimates the number of viable cells in a given cell suspension, how?

Chapter 4

Microorganisms—An Overview

Antonie van Leewenhoek the discoverer of microbial world used the term **Animalcules** for the microscopic organisms. Thus till 1857, the microorganisms were included in the animal kingdom. It was in 1857 when a Swiss botanist Carl Von Nageli proposed a separate class in plant kingdom called **Schizomycetes** to accommodate microorganisms. Later in 1866 Ernst Haekel proposed a third kingdom named **Protista** for microorganisms. The word protista is derived from Greek, which means 'primitive' or 'first'. Rogers Stanier (1957) further divided Protista into two sub-groups on the basis of their cellular organisation. The microorganisms with procaryotic type of cell structure were termed **Lower Protists** and those with eucaryotic type of cell structure were called **Higher Protists**.

The present classification of living beings that is based on cellular organisation and mode of nutrition was proposed by Robert Whittker in 1969. Accordingly there are five kingdoms, however, all the microorganisms are included in three Kingdoms i.e. **Procaryotae**, **Protista** and **Myceteae**.

Kingdom	:	**PROCARYOTAE**
Division I	:	**Gracilicutes**
Division II	:	**Firmicutes**
Division III	:	**Tenericutes**
Division IV	:	**Mendosicutes**
Kingdom	:	**PROTISTA**
Subkingdom I	:	**Protozoa**
Subkingdom II	:	**Algae**
Kingdom	:	**MYCETEAE (FUNGI)**
Division 1	:	**Gymnomycota**
Division 2	:	**Mastigomycota**
Division 3	:	**Amastigomycota**
Kingdom	:	**PLANTAE**
Kingdom	:	**ANIMALIA**

PROTOZOA, ALGAE AND CYANOBACTERIA

In this exercise an attempt will be made to make you familiar with the Protozoans, Algae and Cyanobacteria that are common inhabitants of pond water. Some typical characteristics that differentiate them from one another are discussed below to help you differentiate between the three distinct groups of microorganisms. You will prepare wet mounts and hangings drops of pond water sample and examine microscopically.

Protozoa

Protozoa are unicellular, unpigmented animal like eucaryotic microorganisms belonging to sub kingdom **Protozoa** of Kingdom **Protista**. They are found in soil, water and animal body. Some species are parasites of animals. *Plasmodium*, the malarial parasite is most significant pathogen of humans.

Protozoa are heterotrophs and feed on other microorganisms and on small particulate matter. They lack cell walls and possess distinct nuclei surrounded by nuclear membrane. Some protozoa are covered with a thick cell membrane called **pellicle**. All protozoa reproduce asexually by cell division, however, some possess varying degree of sexuality. On the basis of their mode of locomotion the protozoa are classified into three phylla as described in Table 4.1.

Table 4.1. Distinguishing characters of different Protozoa

Phyla	Sarcomastigophora		Celiophora	Sporozoa
Sub-phyla	Sarcodina	Mastigophora		
Motility	By protoplasmic projections called **Pseudopods**	By a whip-like structure called **Flagella**	By short hair-like structure present all over the body called **Cilia**	Lack locomotory organelles (members are parasites of animals)
Example	*Amoeba*	*Cercomonas*	*Paramecium*	*Plasmodium*

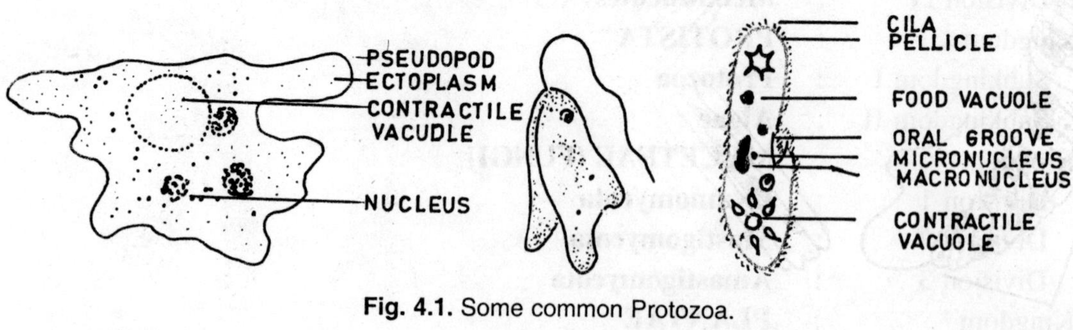

Fig. 4.1. Some common Protozoa.

Algae

Another sub-Kingdom of kingdom **Protista** is **Algae**, which include photosynthetic eucaryotic organisms that do not possess cellular differentiation i.e. they lack true roots, stems and leaves. They are unicellular, colonial as well as multicellular. Algae are found wherever sufficient moisture, favourable temperature and enough sunlight are present. Algae have distinct visible nuclei and chloroplasts. According to the type of pigment, storage food material, chemical composition of cell walls and presence of flagella, algae are classified into seven divisions:

- **Euglenophycophyta (Euglenoids)**
- **Chlorophycophyta (Green Algae)**
- **Chrysophycophyta (Golden brown Algae)**
- **Phaeophycophyta (Brown Algae)**
- **Pyrrophycophyta (Fire Algae)**
- **Cryptomonads**
- **Red algae**

Member of division **Euglenophycophyta** are flagellated and resemble protozoa. Protozoan like characteristics present in euglenophycophyta are (i) the absence of cell wall (ii) presence of a gullet (iii) the ability to ingest food (iv) the ability to assimilate organic substances and (v) the absence of chloroplasts in some species. A majority of algae found in pond water belong to division **Chlorophycophyta**. Golden brown algae included in division **Chrysophycophyta** is a diversified group and includes diatoms also. Cell walls of diatoms are made of pectins, cellulose and silicone. The cell walls are constructed of two halves that fit together like lid and box. The petroleum reserves were probably formulated by accumulation of oil from dead diatoms over millions of years. Members of division **Phaeophycophyta** (Brown Algae) and **Pyrrophycophyta** (Fire Algae) are inhabitants of salt water and are not found in pond water. Some distinguishing characteristics of three divisions of algae generally found in pond water are given in Table 4.2 and representative genera are given in Fig. 4.2.

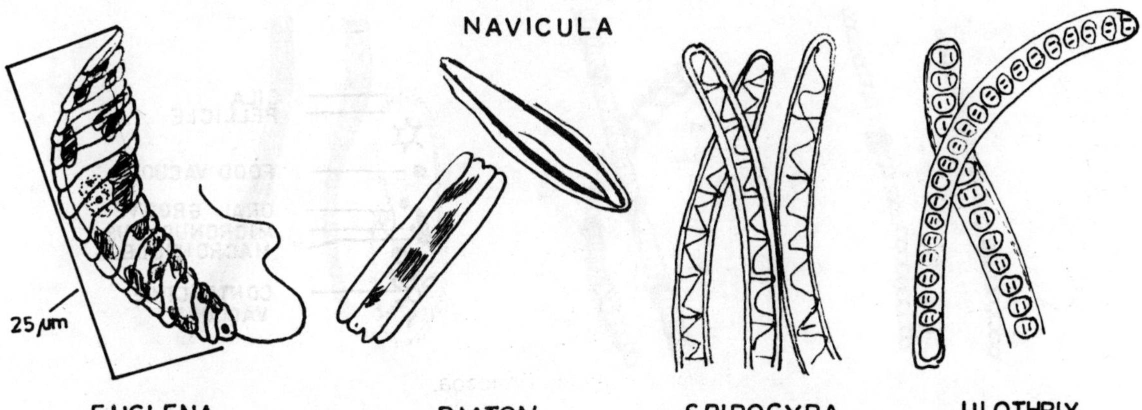

Fig. 4.2. Some common genera of Green Algae found in pond water.

Table 4.2. Distinguishing characters of three Divisions of Algae

Characteristics	Euglenophycophyta (Euglenoids)	Chlorophycophyta (Green Algae)	Chrysophycophyta (Diatoms)
Colour	Possess chlorophyll a and b. Appear green with a red stigma (eye spot).	Possess chlorophyll a and b. Green in colour.	Contain chlorophyll a and c. Members also contain a brownish pigment called *Fucoxanthin*. Appear Golden brown.
Cell arrangement	Unicellular	Unicellular, colonial or filamentous	Unicellular or colonial
Cell wall	Absent	Visible	Visible
Flagella	Present	Present in some species	Absent
Nutrition	Faculative heterotrophic	Autotrophic	Autotrophic
Storage food material	Lipopolysaccharide called **Paramyllum**	Starch	Oils or polysaccharide termed **Leucosin**
Production of oxygen	Present	Present	Present

Cyanobacteria

Cyanobacteria earlier described as **Blue Green Algae** are typical prokaryotic in cellular organisation and belong to Kingdom Procaryotae. Cyanobacteria are unicellular as well as filamentous, autotrophic and fix molecular nitrogen. They vary in colour from black, purple red to green because of varying proportion of different pigments they possess. The pigments in cyanobacteria are not contained in chloroplasts like green algae but are located in granules called **Phycobilisomes**, which are attached to **Thylakoids**. The various types of pigments found in cyanobacteria are **Chlorophyll a**, **Xanthophyll**, **Carotene**, **c-Phycocyanin** and **c-Phycoerythrin**. Some important representatives of cyanobacteria are *Anabaena*, *Nodularia*, *Oscillatoria*, *Calothrix*, *Cylindrospermum* and *Rivularia* (Fig. 4.3).

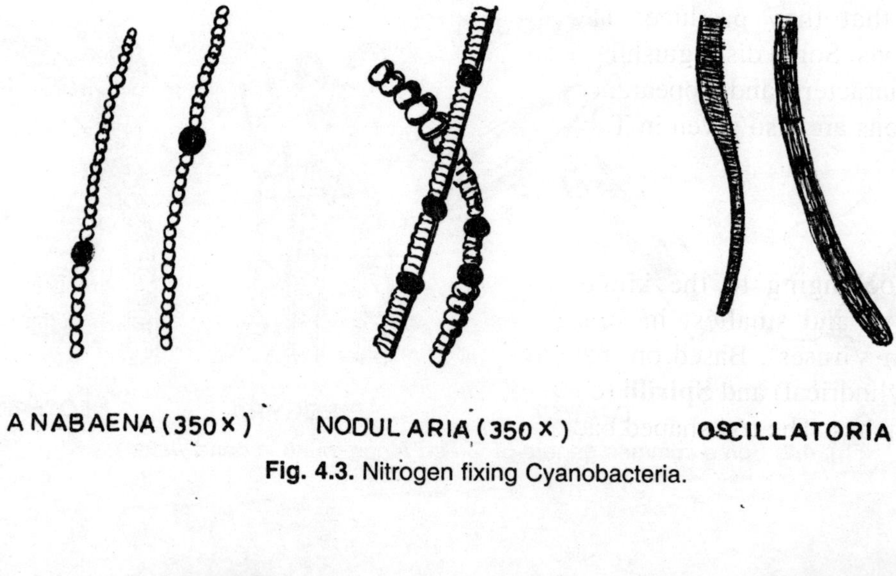

Fig. 4.3. Nitrogen fixing Cyanobacteria.

Fungi: Yeast and Molds

The fungi belong to the kingdom Mycetaeae. Some distinguishing characters of the fungi are i) eukaryotic ii) nonphotosynthetic iii) lack tissue differentiation iv) propogate by asexual and/or sexual spores and v) possess cell walls made of chitin or other polysaccharides. The kingdom Myceteae is divided into three divisions:

Gymnomycota - **Slime molds**
Mastigomycota - **Water molds**
Amastigomycota - **Yeasts, molds, bracket fungi etc.**

The division Amastigomycota includes fungi of economic importance and in this exercise you will try to identify members of this division.

In general, yeasts are unicellular and molds are multicellular. However, many species of fungi are found as yeasts as well as molds. Such fungi are known as **Biphasic**. Yeasts reproduce by budding. A **bud** is an outgrowth of the parent cell also known as **Blastospore**. The bud may separate from the parent cell or remain attached to it (Fig. 4.4).

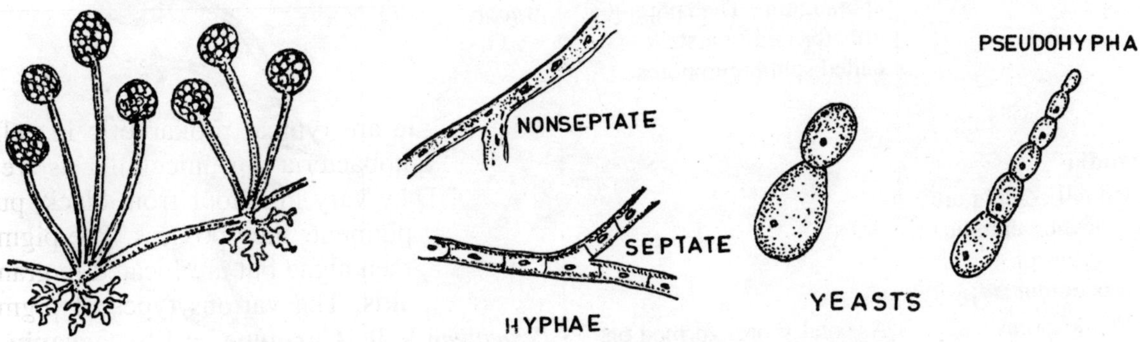

Fig. 4.4. Yeast and Mold.

Molds can be differentiated on the basis of type of hyphae (septate, aseptate) and the type of spore that they produce. The division Amastigomycota is sub-divided into four sub-divisions. Some distinguishing features of these four sub-divisions are listed in Table 4.3. Salient characters and appearances of asexual and sexual spores that differentiate four sub-divisions are also given in Table 4.4.

Bacteria

Bacteria belonging to the kingdom Monera are most widely distributed, simplest in morphology and smallest in size (0.5 to 2 μm) among all the microorganisms known (Exception-Viruses). Based on their morphology bacteria are grouped as **Cocci** (spherical), **Bacilli** (cylindrical) and **Spirilli** (curved). The cocci occur singly, in pairs, as tetrad, in chains and in bunches. The rod shaped bacteria vary in length, diameter and may have square, round

Table 4.3. Distinguishing character of four Sub-Divisions of Amastigomycota

Sub-division	Hyphae	Asexual spore	Sexual spore	Genera
Zygomycotina	Aseptate	Sporangiospores	Zygospores	*Mucor, Rhizopus*
Ascomycotina	Septate	Phialospores	Ascospores	*Penicillium, Saccharomyces cerevisiae*
Basidiomycotina	Septate	None	Basidiospore	*Puccinia*
Deuteromycotina (Fungi imperfactii)	Septate	Conidia, Arthrospore, Blastospore, Chlamydospore	None	*Candida*

Table 4.4. Typess of Asexual and Sexual spores produced by Fungi

Type of spores	Salient characters	Example	Structure
Asexual Spores			
Sporangiospore	Formed inside a sac called sporangium. The sporangia are attached to a stalk called sporangiophores	*Rhizopus, Mucor*	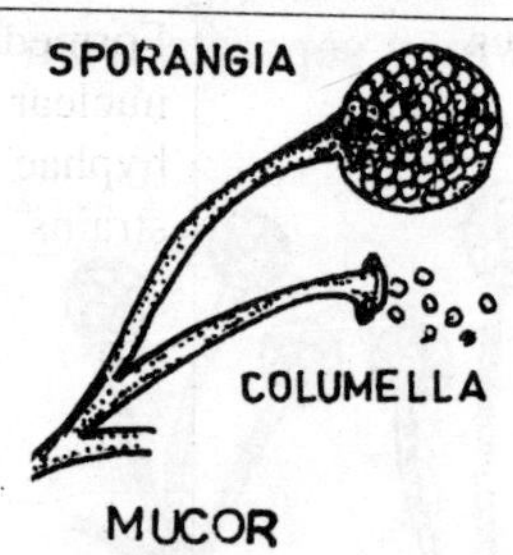
Conidia (a) Small conidia are called Microconidia. Four types of microconidia are:			
1. Phialospores	Asexual spores formed on splcialized hyphae called conidiophores	*Penicillium*	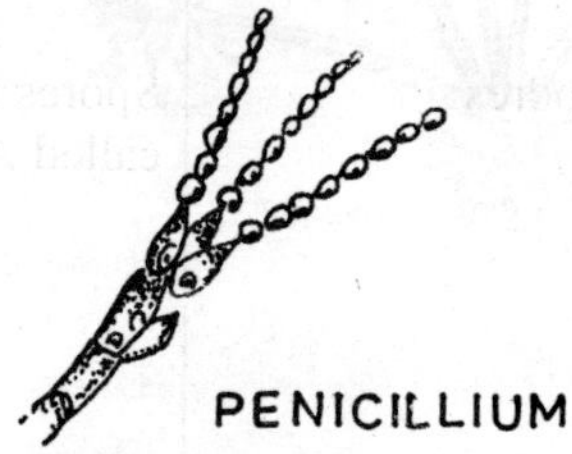
2. Blastoconidia	Conidia produced by vase-shaped cells called phialids	*Cladosporium*	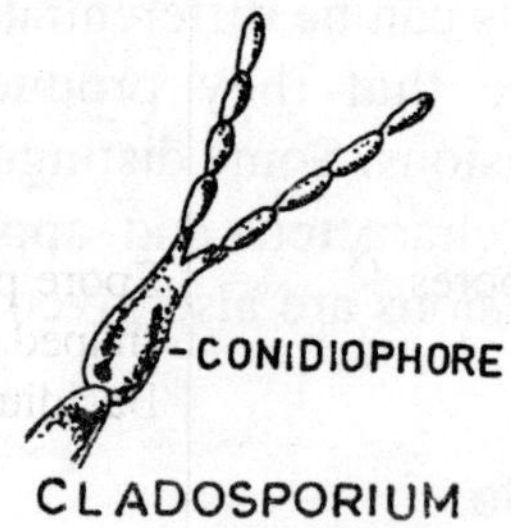
3. Arthrospores	Conidia formed by budding from preexiting conidia	*Oospora*	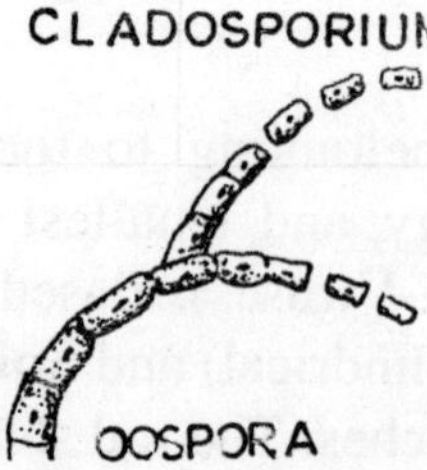

(Continued)

4. Chlamydospores	Conidia formed by separation from hyphal cells	*Candida albicans*	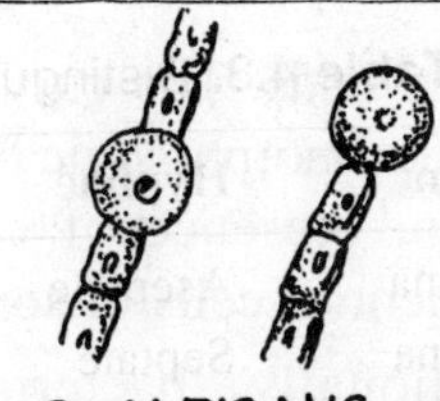
(b) Multicellular, large conidia are called Macroconidia	Conidia formed within or ends of a hypha	*Fusarium, Alternaria*	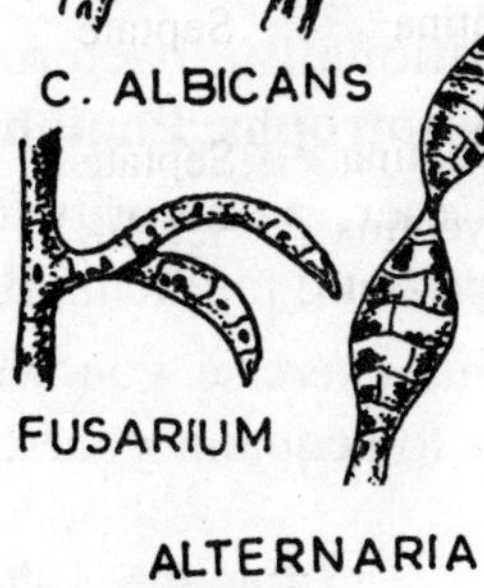
Sexual Spores Zygospores	Formed by fusion of nuclear material from hyphae of two different strains	*Mucor, Rhizopus*	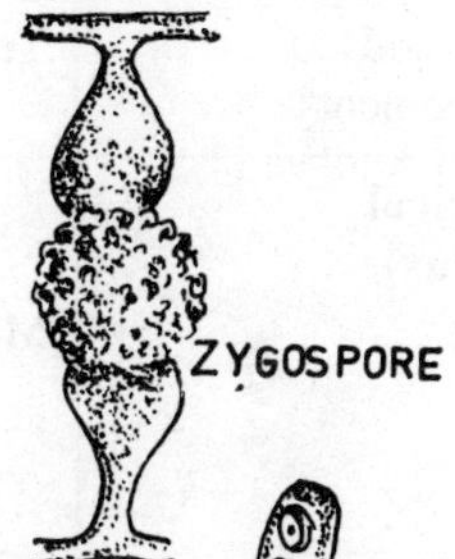
Ascospores	Spores formed in a sac called Ascus (Asci)	*Saccharomyces cerevisiae, Penicillium*	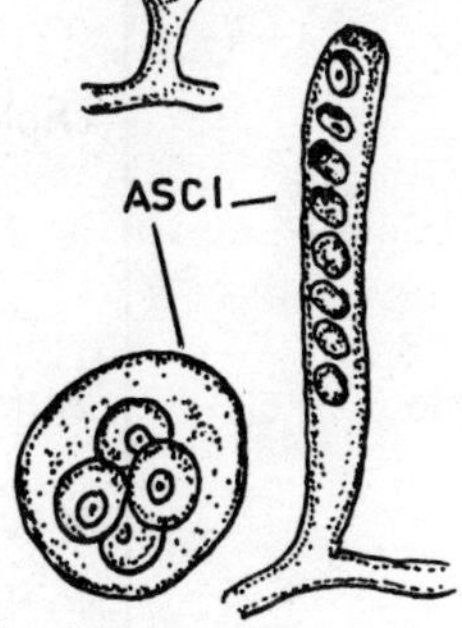
Basidiospores	Spore produced on club shaped body called Basidium	*Puccinia graminis, Agaricus campestris*	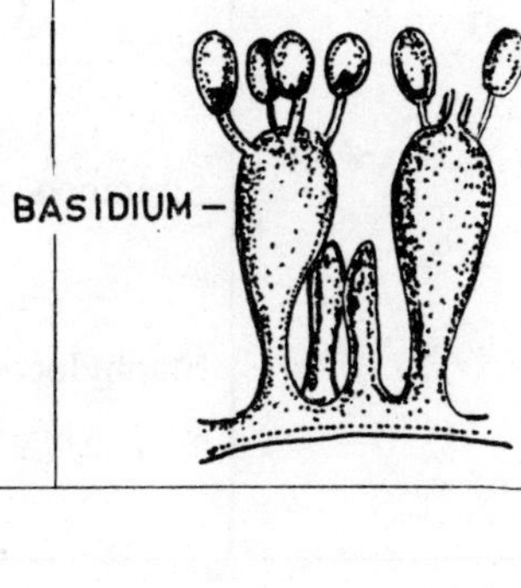

or pointed ends. The curved forms are found as slender, spiral spirochaetes and bent rods. Some other morphological characters of bacteria useful in their identification are endospore formation, presence of capsule, presence and arrangement of flagella etc. Shapes and arrangement of cells of some bacteria are described in Table 4.5.

Nutritionally, bacteria are versatile and belong to all the four categories i.e. **Photoautotrophs**, **Photoheterotrophs**, **Chemoautotrophs** and **Chemoheterotrophs**.

However, a majority of them are chemoheterotrophs. A few bacteria which possess photosynthetic pigments **Bacteriochlorophyll** a,b are photosynthetic.

In this exercise you will prepare wet mounts and perform simple staining of the given cultures to acquaint yourself with different shapes and arrangements of bacteria.

Table 4.5. Different shapes and arrangements of cells among bacteria

Shape and arrangement	Term used	Genus	Appearance
Spherical (Coccus) Single	Monococcus	*Micrococcus* spp.	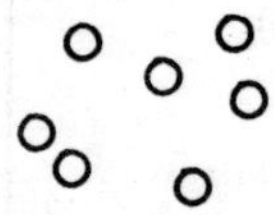
Pairs	Diplococcus	*Neisseria gonorrhoeae*	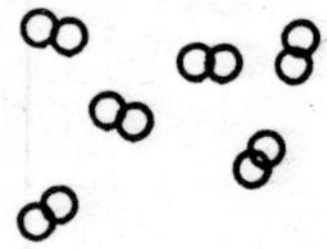
Quadruple	Tetrad	*Pediococcus* spp.	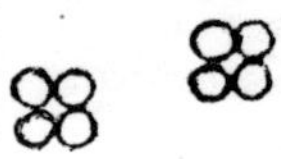
Cuboidal packets	Sarcinae	*Sarcina ventriculii*	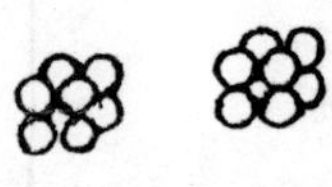
Chains	Streptococcus	*Streptococcus* spp.	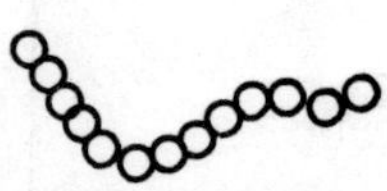
Bunches	Staphylococcus	*Staphyllococcus spp.*	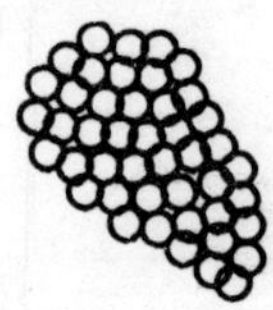

(Continued)

Cylindrical (Bacillus)			
Single	Bacillus	*Bacillus* spp.	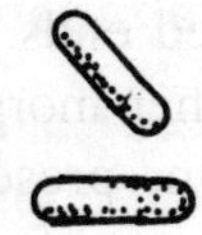
Chains	Streptobacillus	*Lactobacillus* spp.	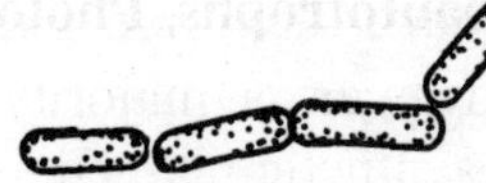
	Pallisade	*Corynebacterium diphtheriae*	
Curved (Spirillium)			
Curved rod	Vibrio	*Vibrio cholerae*	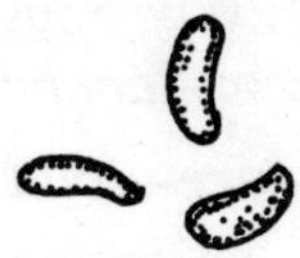
Hellically curved	Spiral	*Spirillum*	
Corkscrew	Spirochaete	*Treponem pallidium*	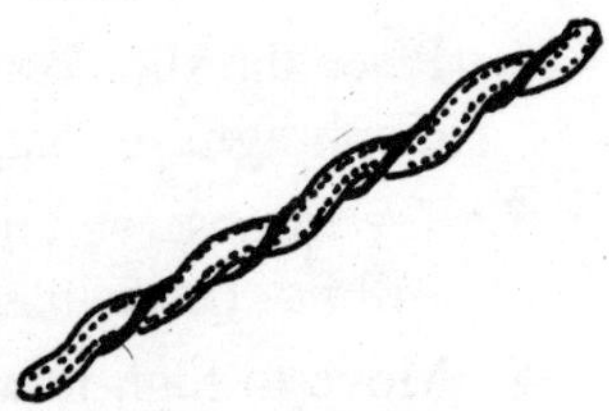
Star formation	Rosette	*Caulobacter* spp.	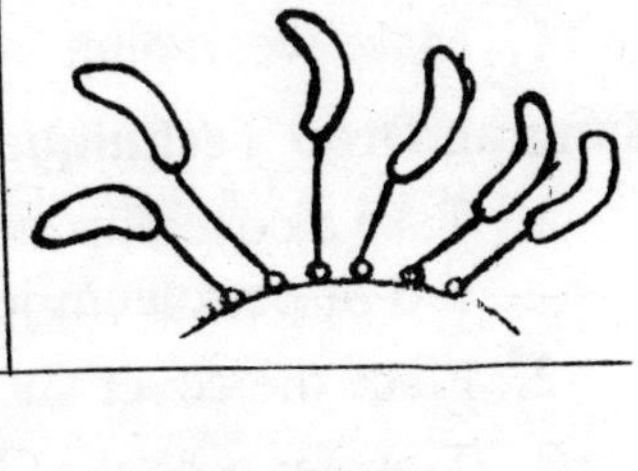

4.1. Examination of Living Microorganisms

The direct examination of live microorganisms is useful in determining size, shape, motility and reaction to various chemicals or anti-sera. Two common and useful methods for direct examination of live microorganisms are Wet mount and hanging drop techniques. Both methods maintain the natural shape of organisms and eliminate the distortion that occur when specimens are dried and fixed.

Objectives:

1. To learn to prepare and observe microorganisms by wet mount and hanging-drop techniques,
2. To compare the shape and size differences among different microorganisms.
3. To distinguish between motility and Brownian movement.

Materials required: Slides, Cover slips, Cavity (depression) slide, Petroleum jelly, Pasteur pipette, Hay infusion incubated in dark, Hay infusion incubated in light, Pond water, Broth culture of *Bacillus* spp. and *Saccharomyces cerevisiae*. Inoculation needle, Microscope.

Procedure—Wet Mount Technique:

1. Shake gently the hay infusion or culture broth.
2. Transfer a drop of hay infusion/culture broth using a sterile Pasteur pipette or inoculation needle aseptically.
3. Flame the lip of the culture tube after removing and before replacing the cotton plug.
4. Do flame and heat the inoculation needle to red hot condition before and after removing a sample.
5. Keep one edge of a cover slip in contact with the culture and lower it gently until it comes in contact with the slide. Ensure that all the air is displaced.
6. Place the slide on the microscopic stage and observe using low power. Adjust the Iris diaphragm so that intensity of light is low.
7. Try to observe larger, more rapidly moving organisms. Bacteria being small in size will not be visible.
8. Move to high power (40×) objective, increase the light intensity and focus. Bacteria shall be visible.
9. Record your observations noting the relative size and shape of the organisms.
10. Some microorganisms are motile while others show Brownian movement.
11. Make wet mount from other samples and observe using low and high power objectives.

Hanging Drop Technique:

1. Take a coverslip and apply petroleum jelly with the help of a tooth pick so that a small rim of petroleum jelly is formed on all the four edges of the cover slip.
2. Place the cover slip on the working table with the petroleum jelly side up.
3. Transfer a drop of hay infusion to the centre of cover slip.

4. Take the cavity slide and place over the drop with cavity side facing downward. Invert the slide quickly so that drop is in suspended position.
5. Examine under low power by locating the edge of the drop.
6. Keeping the light intensity low, observe different types of microorganisms and compare their sizes, shapes and type of movement.
7. Move to high power objective and record your observations.
8. Repeat the above procedure with other samples of hay infusion, pond water or culture broth.

Results and observations: Draw the types of protozoa, algae, fungi and bacteria observed. Indicate their relative sizes and shapes at different magnifications.

4.2. Lacto-Phenol Blue Stain

Lacto-phenol blue stains the fungal cytoplasm as light blue region thus it provides a light blue background against which unstained hyphal walls can easily be seen. Fungal cell being large in size can easily be observed even without staining. However, staining help in observation of other structures of of the fungal cell.

Objectives: To observe the various components of fungal cell.

Materials required: Young cultures of *Aspergillus* sp. and *Penicillium* sp. (3-4 days old)

Procedure:

1. Place a drop of lactophenol blue in the centre of a clean slide.
2. Sterilise the inoculation needle on Bunsen flame and allow it to cool. Transfer a little of fungal mycelium into the lactophenol blue on the slide.
3. Using two mounted needles, tease out the hyphae so that hyphae are separated, which will help in the observation of individual hypha.
4. Cover the preparation with a thin cover slip carefully so that air bubbles are not trapped inside. This can be achieved by placing one edge of coverslip on the slide and lowering the coverslip over the preparation slowly with the help of a mounted needle. In this fashion when needle is removed the coverslip will fall on the preparation with exclusion of air.
5. Examine the slide using low-power and high -dry objectives.

Results: The fungal cytoplasm is seen as lightly stained blue region inside the unstained hyphal cell wall.

Observations: Examine and look for the type of hyphae (septate or aseptate). Also note the type of spore and draw diagrams of hyphae, sporangiophores and spores in your note book.

4.3. Study of Molds by Slide Culutre

Morphology and arrangement of spores is important in identification of fungi. In wet mount preparation of fungal culture, the hyphae and sporangiospores are damaged during transfer with the result the identification becomes very difficult. Slide culture technique is superior to wet mount since hypha, sporangiospores and spore remain intact facilitating identification.

Objectives: To set up a slide culture for study of a fungal culture.

Materials required: Sterile Petri plates, U-shaped glass rods, Sterile glass slide, Cover slips, Petri plates containing Sabouraud's agar, Scalpels, Forceps, Inoculation needle, Sterile water, Sterile filter paper.

Procedure:

1. Aseptically place a sheet of sterile filter paper in the bottom part of a Petri plate using a sterile forceps.
2. Take a sterile U-shaped glass rod and place on the filter paper.
3. Pour enough sterile water into Petri plate to completely moisten it.
4. Using sterile forceps place a sterile glass slide on the U-shaped rod.
5. Cut a piece of 5 mm square block of the medium from the plate of Sabouraud's agar with a sterile scalpel.
6. Pick the block of agar by inserting scalpel into one side and inoculate both top and bottom surfaces of the cube with spores from the mold colony.
7. Place the inoculated block of agar medium in the center of microscopic slide previously placed in the Petri plate. Be sure that one of the inoculated surfaces faces down.
8. Cover the upper surface of agar block with a sterile cover slip. Place the cover of Petri plate and incubate at room temperature for 48 hours.
9. Examine the slide under low power objective. In case growth is inadequate and spores are not visible, incubate the slide for another 48 hours.

Observations: Note the type of hyphae, sporangiophore and spore using low power and high-dry objectives under natural growth conditions of various fungi. Draw diagrams also.

REVIEW QUESTIONS

1. What is the practical value of wet mount and hanging drop techniques?
2. Differentiate between motility and Brownian movement.
3. Why it is difficult to see bacteria in wet preparations?
4. Where from the organisms comes in the infusion?
5. What is the function of petroleum jelly in the hanging drop?
6. Can you distinguish between procaryotic and eucaryotic microorganisms.
7. What are the advantages of using low power objective for observing fungi?

8. Describe some distinguishing characteristics of Sarcomastigophora, Ciliophora and Sporozoa.
9. Define (a) Pseudopods, (b) Pellicle and (c) Eye spot.
10. Write some beneficial and harmful aspects of fungi.
11. Why filter paper is moistened in slide culture technique in the study of fungi?
12. How slide culture technique of fungi is superior to wet mount technique?
13. Write down the industrial significance of yeast.
14. How will you distinguish between a bacterial cell and a yeast cell?
15. Describe in brief the basis of classification of fungi.
16. Write a note on the importance of algae.

Chapter 5

Bacterial Staining

Microorganisms are not only minute in size but are also transparent and colourless when suspended in aqueous medium. Thus visualisation of microorganisms in natural state is difficult. To study the properties of microorganisms and to differentiate them into specific groups for diagnostic purposes microorganisms are stained with different colours. In addition to this, staining enhances contrast between bacteria and the surroundings thus making observations easier.

All bacterial dyes are synthetic aniline dyes. A stain (dye) may be defined as an organic compound containing a benzene ring, a chromophore and an auxochrome group. The chemical group that impart colour to benzene is called chromophore while the chemical group responsible for imparting ionization properties to a chromogen (Benzene + Chromophore) is known as Auxochrome.

$$\text{Benzene} + \text{Chromophore} \rightarrow \underset{\text{(Coloured compound)}}{\text{Chromogen}} + \text{Auxochrome} \rightarrow \text{Stain}$$

The chromogens though coloured compounds but are generally insoluble in water and do not have staining properties i.e. the colours don't stick. However, salts of these compounds are more soluble, have better penetration and stain more permanently. Thus they are the true dyes. A stain binds to a macromolecular cellular component (protein or nucleic acid) depending on the electrical charge found on the chromogen portion of the stain and on the cellular component to be stained. Dyes are of two types Acidic and the Basic.

Acidic Dyes: These are anionic i.e. on ionization the chromogen portion exhibits negative charge and thus has greater affinity for the positive constituents of the cell. Proteins are positively charged cellular component which bind to negatively charged, anionic, chromogen of an acid stain. Thus acidic stains owe their coloured properties to the anions and are generally sodium, potassium, calcium or ammonium salts of coloured acids e.g.. Picric acid.

Basic Dyes: Basic dyes are cationic i.e. upon ionization the chromogen portion exhibits a positive charge that has a strong affinity for the negatively charged constituents of the cell. Nucleic acids are negatively charged and readily bind with the coloured positively charged cationic chromogen of a basic stain. The basic stains owe their colour properties to cations. Basic dyes are generally chloride or sulfate salts of coloured bases. Methylene blue is an example of basic stain.

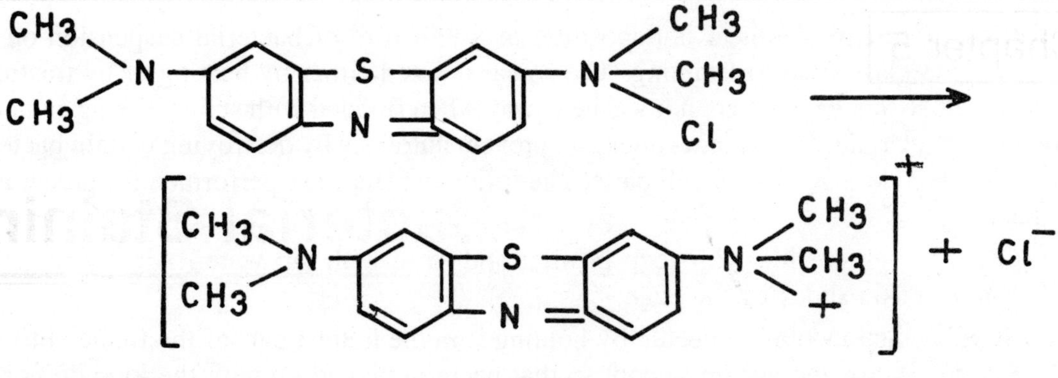

Methylene Blue

Types of Staining

Numerous staining techniques are known for the study of morphological properties and staining of cellular structures. All these techniques may be divided into three groups based on the staining procedure and their purposes.

Simple staining: In simple staining only one stain is used. This type of staining is useful for visualisation of shape (cocci, bacilli or spirilli) and arrangement (single, pairs, chains or clusters) of cells. Negative staining in which the background is stained leaving the bacteria unstained may also be included in simple staining.

Differential staining: This is the type of staining in which two contrasting stains are used to differentiate between two distinct groups of bacteria e.g. Gram stain and Acid fast stain.

Structural staining: This is a type of differential staining in which different structures of a cell can be stained for better visualisation e.g. Spore stain, Capsule stain, Flagella stain and Nuclear stain.

5.1. Simple Staining

The cytoplasm of bacteria possesses a strong affinity for stains when treated with methylene blue (a basic dye). The bacterial cells are stained blue uniformly since the cytoplasm combines with the stain and all structures within the cell are concealed under the blue stained cytoplasm.

Objectives:

1. To learn to make and fix a smear.
2. To learn the basic mechanism of staining.
3. To perform a simple staining.

Material required: Clean slides, Broth cultures of *Bacillus* spp. and *Micrococcus* spp., Methylene blue solution" .

Procedure:

Preparation of smear: A smear is deposition of a thin film of bacterial suspension on the surface of clean glass slide for staining. The smear is fixed either by heating or by treatment with chemicals so that bacteria are not washed away when flooded with stains. Fixing enhances adherence of bacterial cells to glass slides and prevent autolysis by destroying certain bacterial enzymes irvolved in digestion of cell parts. The following steps are performed for preparation of a smear.

1. Clean the glass slide with soap powder and rinse with tap water. Air-dry the slide. Handle clean slides by the edge.
2. Sterilise the inoculation needle by holding it in the hottest part of the flame until it is red-hot. Allow the needle to cool so that bacteria picked up with the loop do not get killed.
3. Take 1 or 2 looful of bacterial suspension prepared earlier and place in the center of the slide.
4. Spread the bacterial suspension in approximately 1 cm^2 with the help of needle. Keep the slide to let the smear dry. Do not blow on the slide. Do not flame or heat the slide to dry the smear as flaming will distort the cell's shape.
5. Hold the slide by the edge and fix the smear by passing the slide quickly through the blue flame 2-3 times. Alternately cover the smear with 95 per cent methyl alcohol for 1 minute. Let the alcohol run off and dry the slide before staining.

Staining procedure:

1. Prepare thin smears of both the cultures. Air dry and heat fix them.
2. Flood the smears with Methylene blue and allow to stand for one minute.
3. Wash the slides to remove the free stain in gently running tap water.
4. Blot dry the slides with a piece of clean blotting paper.
5. Keep a drop of immersion oil on each of the smear and observe using oil immersion lens of the microscope.

Results: The bacteria stain deep blue with Methylene blud. While the cells of *Bacillus* sp. are rod shaped (cylindrical) the cells of *Micrococcus* spp. are coccoid.

Observations: Describe the morphology of the organisms with respect to their shape and arrangement of cells.

5.2. Negative Staining

In negative staining while the bacteria remain colourless, the background is stained. Because both, the bacteria and stain used have negative charges, the bacteria are not stained because of ionic repulsion. Since in this staining technique heat or chemicals are not used for fixation, the morphology of bacteria is not distorted. Thus the negative staining is very useful in the study of cell size and morphology.

Objectives:

1. To learn and prepare a negative stain.
2. To interpret the application and mechanism of negative staining technique.

Materials required: Clean slides, Sterile tooth picks, Nigrosine black dye. 24 hour old culture of *Bacillus subtilis* and *Staphyllococcus* spp.

Procedure:

1. Clean the slide greese free with soap powder, rinse and dry.
2. Place a drop of nigrosine black at one end of the slide.
3. Take a loopful of broth culture with a sterile inoculation needle and mix into the nigrosine drop. Do not spread the drop or let it dry.
4. Now spread the drop with the end edge of another slide to produce a smear of varying opacity.
5. Let the smear air dry and examine microscopically using low, high dry and oil-immersion objectives.
6. Prepare another slide by scrapping the base of your teeth and gums with a sterile toothpick.
7. Examine microscopically.

Results: In negative staining bacteria appear colourless against a black background.

Observation: Describe the microscopic appearance of different types of bacteria you have observed.

5.3. Gram Staining

The Gram stain is one of the most important techniques used in bacteriology. In 1884 a Danish scientist named Christian Gram observed that bacteria when stained with aniline-gentian violet and treated with Lugol's iodine could be divided into two groups on the basis of their behaviour with absolute alcohol. While one group retained the violet dye, the second group got decolourised with alcohol treatment. The decolourised cells are stained with a contrast counter stain usually safranin. Bacteria resistant to decolourization will retain violet colour are referred as Gram positive. Microorganisms unable to retain crystal violet stain will take the counterstain and thus exhibit a pink colour. Such microorganisms are termed as Gram negative.

Many investigators have tried to determine the mechanism of Gram staining. An explanation based on permeability differences to decolourising solvent has been most favourable and accepted universally. Accordingly Lugol's iodine act as a mordant-a substance that forms an insoluble complex by binding to primary stain (crystal violet). Crystal violet-iodine (CV-I) complex binds to magnesium ribonucleic acid component of cell wall forming a tertiary complex. Gram-negative bacteria have more lipid contents, which facilitate faster movement of decolourising agent (absolute alcohol) compared to Gram-positive bacteria. The peptidoglycan of Gram positive bacteria which is thicker than that of Gram negative bacteria also resist penetration of decolourising agent. Thus absolute alcohol penetrate faster inside the Gram negative bacteria and break the tertiary complex formed by cell wall component and CV-I complex.

In fact the basis of this differentiation is one of the rate, rather than an absolute characteristic of bacteria. If treated with alcohol for long even the Gram positive bacteria will also get decolourised. On the other hand, insufficient treatment with alcohol will result in retention of crystal violet by Gram negative cells. Thus the staining by this method need expertise and shall be performed carefully. Secondly Gram staining is performed on young cultures (24 to 48 hour old) since many bacteria rapidly lose their ability to retain the violet stain as they grow older and therefore, appear as Gram negative.

Objectives:

1. To learn the rationale and mechanism of Gram stain.
2. To perform and interpret Gram stain.
3. To differentiate between Gram positive and Gram negative reactions.

Materials required: Clean slides, Young broth cultures of *Bacillus* spp. and *Escherichia coli*, Crystal violet - ammonium oxalate solution, Lugol's iodine. Iodine-acetone/Acetone alcohol solution, Safranin solution.

Procedure:

1. Prepare thin smears of both the cultures on separate slides. Air dry and heat fix.
2. Cover the smears with crystal violet-ammonium oxalate solution and allow to stand for one minute.
3. Wash with gently running tap water.
4. Rinse the slides with Lugol's iodine reagent and then apply the same reagent for 30 seconds.
5. Apply alcohol-acetone reagent drop wise until the acetone solution flows off the smear free from crystal violet.
6. Wash the slides immediately with tap water to avoid excess reaction of alcohol-acetone reagent.
7. Cover the smears with safranin reagent and keep for 30-60 seconds.
8. Wash the slides as above, blot dry the slides. Do not rub.
9. Examine the smears using high dry and oil immersion lens of microscope.

Results: Bacterial cells stained violets are described as Gram positive and those appearing pink are termed Gram negative. Cells of *Bacillus* spp. are large rods (cylindrical) and stain violet, whereas cells of *E. coli* are small rods and stain pink.

Observations: Describe and draw the shape, size and arrangement of Gram positive and Gram negative bacteria.

5.4. Acid-Fast Staining

Paul Ehrlich in 1882 discovered that the tubercle bacilli, *Mycobacterium tuberculosis* retained the primary stain, carbolfuchsin even after washing with an acid-alcohol mixture. Later it was found that most bacteria are decolourised by acid-alcohol, however, members of families Mycobacteriaceae and Nocardiaceae are acid fast and retain the primary stain after acid-alcohol treatment. Since both the genera *Mycobacterium* and *Nocardia* contain pathogenic species, the acid-fast staining has become one of the most important diagnostic procedures. The cell walls of acid fast organisms contain a wax like lipid, mycolic acid that renders the cell wall of these bacteria impermeable to most stains. Thus the smear has to be heated to facilitate penetration of the stain into the bacterial cell envelope. The technique employed today was developed by Franz Ziehl and Friedrich Neelsen and is known as Ziehl-Neelsen procedure. This procedure was modified by Kinyoun who increased the concentrations of phenol carbolfuchsin to eliminate heating.

The principle of acid-fast staining is that the fuchsin is more soluble in carbolic acid (phenol) than in water and carbolic acid is more soluble in lipid (mycolic acid) than in acid-alcohol mixture. Thus carbolfuchsin has a higher affinity for lipid than acid-alcohol and is retained by the cell when washed with acid-alcohol.

Objectives:

1. To perform and interpret acid fast stain.
2. To interpret the mechanism underlying acid fast staining.

Materials required: Clean slides, Cultures of *Mycobacterium phlei* and *Staphylococcus aureus*. Electric hot plate, Beaker, Carbolfuchsin (or Kinyoun's carbolfuchsin), Acid-alcohol, Methylene blue.

Procedure:

1. Prepare smears of both the cultures. Air dry and heat fix by passing over the flame.
2. Cover the smear with carbolfuchsin keeping the slide over boiling water for 5 minutes. If the stain boils, add more stain. Heating over steam is not needed if Kinyoun's carbolfuchsin is used.
3. Rinse the slide gently with distilled water using a wash bottle.
4. Wash the smear with decolourizing reagent (acid-alcohol mixture) for 20-30 seconds or until no more red colour run off the slide.
5. Rinse with distilled water to stop the action of acid-alcohol mixture.
6. Counter stain with methylene blue for 1 minute.
7. Rinse with distilled water to remove excess of methylene blue.
8. Blot dry and examine microscopically and record your observations.

Results: The cells which retain deep red colour of carbolfuchsin are Acid-fast while those which are stained blue with counter stain are Acid slow.

Observations: Draw the morphology and record the colour of the bacteria to determine acid-fast reaction.

5.5. Spore Staining

Spore forming bacteria constitute a unique group of bacteria and have been described in separate section in Bergey's Manual of Systematic Bacteriology. It was thought for many years that sporulation occurred when the cell was encountered with starvation or a toxic environment. However, today we know that nutritional and environmental requirements for sporulation are not too different from those required for vegetative cell. Thus, the process is a mode of differentiation and just a part of life cycle of some bacteria. Nevertheless, sporulation helps in the survival of those organisms, which produce them.

Though many bacteria produce spores, two genera *Bacillus* (aerobic) and *Clostridium* (anaerobic) are important. Both of them form spores inside the cells and thus the spores are known as **Endospores**. Important features of endospores are resistance to heat, dessication and toxic substances. Several layers (spore coat) surround the mature endospore and its cytoplasm is extremely dehydrated. These properties make the endospore impermeable to stains. Thus spore cannot be stained by ordinary methods like simple stain or Gram stain. When spore-bearing cells are stained by this method, the cells are stained whereas the spores appear colourless. To enable the stain to penetrate into the spores, the spores are given heat or chemical treatments. Once stained, spores will resist replacement with counter stain safranin.

Objectives:

1. To prepare and interpret endospore stain.
2. To learn to detect the presence of bacterial endospore in a culture.

A. Hot spore staining (Schaeffer-Fulton Method)

Materials required: Clean slides, 24 hour old slant cultures of *Bacillus subtilis* and *Staphylococcus aureus*, Malachite green, Safranin.

Procedure:

1. Prepare smears of the cultures provided to you separately. Air dry and heat fix.
2. Rest the slides on two sides of a tripod. Flood the smears with Malachite green solution.
3. Heat the slides from undersurface until stain begins to steam. Continue steaming for five minutes. Do not allow the stain to dry by adding fresh stain to the smear.
4. Allow the slide to cool and wash with slowly running tap water.
5. Counterstain with Safranin for one minute.
6. Rinse the slide with water, blot dry and examine under oil immersion.

Results: The spores are stained by Malachite green and appear green. Vegetative cells appear pink. A fast green endospore can be seen inside cylindrical pink coloured *Bacillus*. *Staphylococcus* does not form spore and is stained pink.

Observations: Record the shape and colour of vegetative cell. Also record the shape, colour and position of spore.

B. Cold Spore Staining

Another method of staining spores is by the use of strong heat on smear and strong acid. The stain applied is cold.

Materials required: Clean slides, Solution of 12 N Hydrochloric acid or 12N Nitric acid. Solution of 1N Sodium hydroxide. Crystal violet-phenol-ethanol solution. 24 hour old slant culture of *Bacillus* spp.

Procedure:

1. Prepare a smear of given culture. Fix in blue tip of Bunsen burner flame for six seconds. Allow it to cool.
2. Apply hydrochloric acid solution for five seconds.
3. Wash in slowly flowing tap water and neutralize the acid with sodium hydroxide solution for 30 seconds.
4. Rinse with water
5. Apply crystal violet solution for two minutes.
6. Wash and blot dry in between two pieces of blotting paper.
7. Examine the slide using oil-immersion objective.

Results: Spores are stained deep purple which can be seen inside the weak purple vegetative cells.

Observations: Record the shape and colour of vegetative cell and spore. Note the position of spore inside the cell.

5.6. Capsule Staining

Many bacteria secrete inert (uncharged), water soluble polysaccharides that adhere to their surface forming a viscous coat. When these polysaccharides are present uniformly in round or oval shape around the cell surface it is called **Capsule**. When the polysaccharides are irregular and loosely bound to the bacterium it is known as **Slime layer**. Because of non-ionic nature, stains do not adhere to capsules. Thus in capsule staining, the bacteria and the background are stained leaving the capsule unstained. In this sense it is a type of negative staining.

Objectives:

1. To learn to perform capsule staining.
2. To differentiate between bacterial capsule and artifects.

Materials required: Clean slides, Congo red, Acid alcohol, Acid fuchsin, 24 hour cultures of *Azotobacter chrococcum,* and *Enterobacter aerogenes.*

Procedure:

1. Place a drop of congo red in the centre of a clean slide.
2. Pick a small amount of culture from the Petri plate or slant culture with the help of a sterile inoculation needle.
3. Mix the culture with congo red and prepare a thin smear.
4. Fix the smear with acid alcohol for 15 seconds.
5. Wash the slide with distilled water and cover the smear with acid fuchsin for 1 minute.
6. Wash with distilled water and blot dry. Examine microscopically using high dry and oil immersion objective.

Results: The bacteria will stain red and the capsule will look colourless against a dark blue background.

Observations: Record the comparative size of the capsule (small, moderate, large).

5.7. Flagella Staining

Flagella, responsible for motility in bacteria are thin proteinaceous structures that originate from the cytoplasm and are projected out through the cell envelop. The flagella are fragile and are removed even by mild shaking of broth cultures. The bacterial flagella range from 10-20 μm in length though in some bacteria it may be as long as 70 μm. The diameter of flagella is approximately 20 nm. Because of lesser diameter flagella are not visible with a light microscope and need to be stained after coating with a mordant, which increases their diameter.

Objectives:

1. To prepare and interpret flagella stain.
2. To recognize different types of flagellar arrangements.

Materials required: Clean slides, Flagella mordant, Ziehl's carbolfuchsin, 18 hour old culture of *Proteus vulgaris* on solid agar medium.

Procedure:

1. Culture to be used for flagella staining needs special care to avoid damage to flagella. Because of this the culture of *Proteus vulgaris* is grown on the solid surface of agar media in a Petri plate.
2. Carefully without touching the growth cut a piece of agar along with growth of *Proteus* with the help of sterile scalpel.
3. Using a forcep pick and place the piece with culture side down on a clean glass slide.
4. Remove the piece of agar and allow the culture adhering to the slide to air dry.
5. Flood the slide with flagella mordant and let it stand for 10 minutes.
6. Gently rinse with distilled water.

7. Cover the slide with Ziehl's carbolfuchsin for 5 minutes.
8. Rinse with distilled water, air dry (blot dry not desirable) and examine microscopically.

Results: Both the bacteria and flagella stain dark red and can be visualized against a colourless background.

Observations: Record your observations with regard to position of flagella (polar, peritrichous etc.).

5.8. Viability Stain (Modified Proca-Keyser Method)

The viability staining helps in distinguishing between living and dead cells. It is based on the fact that certain changes occur in dead cells, which give different response to stains as compared to live cells.

Objectives: To distinguish between dead and live cells

Materials required: Old culture of *Bacillus* spp., Loeffler's methylene blue, Dilute carbol fuchsin, Glass slides.

Procedure:

1. Prepare a smear of the culture and heat fix.
2. Stain with Loeffler's methylene blue for 10 minutes.
3. Rinse with tap water until the smear becomes pale blue.
4. Run dilute solution of carbol fuchsin down the slide for a short while.
5. Wash the slide immediately and blot dry.
6. Examine using oil immersion lens.

Results:

1. Living bacteria stain purple while dead cells stain red or pink.
2. Live spores stain faintly pink or mauve and dead spores look blue

Observations: Examine the slide under oil immersion lens and count the number of dead and live cells/spores.

REVIEW QUESTIONS

1. What is the purpose of heat/chemical fixation? What happens if too much heat is applied?
2. Why are basic dyes more effective for bacterial staining than acidic dyes?
3. What are the advantages of negative staining?
4. Why do the bacteria remain colourless in negative staining?
5. Why methylene blue/crystal violet can not be used in place of nigrosine in negative staining?

6. Why the size of cell is considered to be more accurate in negative stain than in simple staining.
7. Which microscopic technique gives a field similar in appearance to that seen in the negative staining?
8. India ink can give the appearance of negative staining when used in wet mount. Why? (India ink is a suspension of 0.5-1.0 μm carbon particles in water)?
9. Distinguish between simple, differential and structural staining?
10. Explain the functions of the following in differential staining: (i) Primary stain, (ii) Counter stain, (iii) Mordant.
11. Write the principle underlying Gram stain.
12. Can iodine be applied before primary stain in Gram stain?
13. Why old Gram-positive cells stain Gram negative?
14. Why heat treatment is essential while applying primary stain in acid-fast staining?
15. Write down the function of acid-alcohol treatment in acid-fast staining. Why can not we use ethyl alcohol in place of acid-alcohol?
16. Name some organisms which are identified by acid-fast staining.
17. What diseases are diagnosed using the acid fast staining.
18. How does the acid fast stain related to Gram stain?
19. Why heating is necessary in spore staining?
20. Endospore can also be seen by simple staining. Why then endospore staining is used?
21. Write the function of washing with tap water in spore staining?
22. Describe the microscopic appearance of spore and vegetative cell in Dorner's endospore stain.
23. Write down the significance of capsule.
24. While staining capsule, heat fixation of smear is omitted, why?

Chapter 6

Culture Techniques

To cultivate microorganisms in the laboratory, we must fulfil their nutritional requirements. Like all other living beings, microorganisms need (i) water, (ii) a source of carbon, (iii) an energy source, and (iv) some inorganic compounds. These substances which are required for the growth of microorganisms are termed **Nutrients**. A combination (solution) of nutrients prepared in the laboratory for the growth of microorganisms is called a **Culture medium**.

A majority of microorganisms have simple nutritional requirements and they can be cultivated on culture medium containing sugars, salts and water without much difficulty. However, some microorganisms require certain unusual nutrients termed as **Growth factors**. These needs make it difficult or sometimes impossible to grow them in the laboratory.

Nutritional Needs of Bacteria

Any medium that is to be suitable for a specific group of organisms must take into account the following seven factors i.e. water, carbon, energy, nitrogen, minerals, growth factors and pH.

Water: Water is one of the most important and essential nutrients. All biochemical reactions within the cell require adequate amount of water. Water accounts for 70-80% of total weight of microbial cell. In unicellular organisms water molecules pass freely in and out of the cell and help in the transportation of nutrients inward and secretion of metabolites outward. Since the presence of salts in natural water affect the quality of media, it is always better to use distilled water in the preparation of media.

Carbon: Carbon is the constituent of all organic cell materials. All organisms have been categorized into two groups according to their source of carbon. **Autotrophs** are those organisms that can obtain the carbon from carbon dioxide for the synthesis of cell materials. On the other hand, those organisms that need organic sources of carbon are termed as **Heterotrophs**. Since CO_2 is present abundantly in the atmosphere, there is no need to provide a source of carbon in the medium for the growth of autotrophs.

Energy: Energy is required for biosynthesis of building blocks, polymerization and assembly reactions, transport of solutes and motility of the microorganisms. Two major sources of energy for microorganisms are solar energy and chemical energy. Some microorganisms have pigments and thus are capable of utilizing solar energy. Such microorganisms are termed **Phototrophs**. However, a majority of microorganisms require organic/inorganic compounds to fulfil their energy requirements. Such microorganisms are called **Chemotrophs**. Thus for phototrophs there is no need to include components for providing energy, whereas, inorganic/organic compounds have to be added in the medium for the growth of chemotrophs. Mostly an organic compound provides both carbon as well as energy for chemotrophs.

Nitrogen: Nitrogen another mineral nutrient required in large quantity is a constituent of proteins, nucleic acids and co-enzymes. Except some bacteria which are capable of fixation of atmospheric molecular nitrogen, all others have to be provided nitrogen in the form of inorganic or organic compounds. While heterotrophs can get their nitrogen requirements fulfilled from amino acids, peptides, peptone, beef extract etc., autotrophs have to be provided with inorganic nitrogen.

Minerals: Several inorganic nutrients required for the growth of microorganisms are phosphorus, sulphur, sodium, potassium, calcium, magnesium, manganese, iron, zinc, copper and cobalt. These are required in small quantities and are supplied in the form of salts.

Growth factors: Some microorganisms are unable to synthesize one or more essential component of the cell which are termed **Growth factors**. This may be an amino acid, a purine or pyrimidine base or a vitamin. Thus these growth factors have to be included in the medium. Beef extract or yeast extract serves as a source of growth factors for many heterotrophic microorganisms. Many pathogens require enriched media supplemented with animal blood.

Hydrogen ion concentration (pH): Most of the bacteria grow best around 7 or slightly lower pH. Thus pH of the nutrient broth should be adjusted to 6.8. Pathogens usually prefer an alkaline pH. Trypticase soy broth, a medium for pathogens is thus adjusted to pH 7.3. Fungi prefer acidic conditions (pH 5-5.5) and actinomycetes grow best under alkaline pH (8-9).

Defined Versus Complex Media

A culture medium where all the chemicals and their concentrations are known is referred to as a **Synthetic** or **Chemically defined** medium (Tables 6.1 and 6.2). In contrast a **Complex medium** is where exact chemical constituents and their concentrations are not exactly known. Such media contain beef extract, yeast extract or fruit/vegetable juice etc. to supply water-soluble vitamins and many other growth factors. Complex media are easier to prepare compared to synthetic media and are capable to support the growth of most microorganisms. On the other hand, the synthetic media are easily reproducible.

Special Media

Selective media: The media that allow the growth of certain types of microorganisms is termed as selective media. This is achieved by

Table 6.1. Composition of minimal agar (Synthetic medium)

Component	Amount (g)	Purpose
Glucose	2.0	Source of carbon and energy
Ammonium sulphate	1.0	Source of nitrogen
Dipottasium hydrogen orthophosphate	7.0	Source of phosphorus
Magnesium sulphate	0.5	Source of sulphur
Agar	15.0	Solidifying agent
Distilled water	1 litre	Solvent and source of trace

Table 6.2. Composition of nutrient agar (Complex medium)

Component	Amount (g)	Purpose
Peptone	5.0	Source of carbon and energy
Beef extract	3.0	Source of vitamins and salts
Sodium chloride	8.0	Source of salts
Agar	15.0	Solidifying agent
Water	1 liter	Solvent and source of trace elements

(i) Excluding a critical nutrient that make it unfavourable for most but not all microorganisms. e.g. Jensen's nitrogen free agar medium of *Azotobacter*.

(ii) Addition of some inhibitory substance that prevents growth of certain type of microorganism is added into the medium e.g. Penicillin is addad to prevent growth of Gram positive bacteria.

Differential media: Such media contain certain substances (e.g. dyes) that differentiate one species from another. One such example is Eosin methylene blue agar, which is used to differentiate *Escherichia coli* from other enteric group of bacteria.

A microbiological medium can be prepared in three consistencies i.e. liquid, solid and semisolid. **Liquid media** are required for the cultivation of microorganisms in large quantity, fermentation studies and many other biochemical tests. **Solid media** are prepared by adding solidifying agent such as agar, gelatin or silica gel to a liquid medium. A good solidifying agent is one that is not utilized by microorganisms, remain in solid state at room temperature and is not inhibitory to any microorganism. Agar is one of the most widely used solidifying agent. Solid media are required for isolation of microorganisms, standard plate count etc. In **Semisolid medium**, the solidifying agents (agar or gelatin) are added in lesser quantity so that it appears jelly like and does not solidify. It is useful in the study of motility.

6.1. Preparation of Media

Microorganisms show a considerable variation in their habitat and nutritional requirements.

Thus no single medium can support the growth of all microorganisms. Therefore, a medium for any particular species must be selected carefully. Four basic steps are involved in the preparation of media irrespective of the type of media (i) preparation of solution of the chemicals, (ii) adjustment of pH, (iii) dispensing the media, and (iv) sterilisation. In this exercise you will prepare a synthetic and a complex medium.

Objectives:

1. To prepare different types of media for the growth of different bacteria.
2. To interpret the principles of microbial nutrition.

Materials required: Peptone, Beef extract, Sodium chloride, Glucose, Ammonium sulphate, Potassium dihydrogen orthophosphate, Magnesium sulphate, Agar, pH meter, Dilute NaOH and Dilute HCl.

Procedure:

1. Weigh the components of the two media separately on a balance as per quantities given in Tables 6.1 and 6.2.
2. Prepare a solution of each media by dissolving the appropriate amounts of each component in distilled water.
3. Check the pH using a pH meter and adjust if needed by the addition of dilute NaOH or HCl.
4. The broth thus prepared is dispensed into clean glass test tubes or Erlenmeyer flasks.
5. Tubes and flasks are plugged with non-absorbent cotton, the tubes are placed into wire or aluminium baskets and covered with grease proof paper or aluminium foil to prevent wetting of cotton plugs.
6. To prepare a solid medium, the required amount of agar is added before sterilisation.
7. The agar medium is required for preparation of agar slants or plates.
8. To prepare slants, the agar in the medium is melted in an autoclave and 5 ml of the medium is dispensed in glass test tubes. The tubes are plugged with cotton and sterilised.
9. To prepare plates, the agar medium is prepared in 250 ml or 500 ml flasks. The medium is sterilised and maintained in the molten state by keeping the flasks in a water bath at 50°C. When needed, the agar medium is poured into pre-sterilised plates under aseptic conditions and allowed to stand till the agar solidifies.

Precautions:

1. To prepare solution, various components of the medium should be added one by one to prevent precipitation.
2. Agar should not be melted on a flame to prevent charring.
3. Agar medium when prepared in flasks, the flasks should not be filled more than half. This will prevent the wetting of plugs with medium during sterilisation.

6.2. Aseptic Technique

Microorganisms are cultured in the laboratory to facilitate identification and to study their cultural and metabolic properties. Thus microorganisms are **Inoculated** (introduced) into culture media in a manner that unwanted microbes or **Contaminants** do not get entry into the media. The technique used to exclude contaminants is termed as **Aseptic technique**.

All culture media are sterilised or rendered free of all types of life prior to use. Containers (test tube or Petri plates) of culture media should not be left open before or after the transfer work is over. Transfer of microorganisms is performed with a sterile **Inoculation needle/loop**. Inoculation needle is a heat resistant, noncorroding nichrome wire attached to a glass rod or an insulated handle. In some cases cultures may also be transferred with sterile cotton swabs, pipettes, syringes or tooth picks depending upon the applications.

Objectives:

1. To understand the rationale for aseptic technique.
2. To learn to transfer microorganisms under aseptic conditions.

Materials required: Tubes containing nutrient broth, Nutrient agar slants, Nutrient agar deeps, Inoculation needle, *Bacillus subtilis* broth, *Escherichia coli* broth, *Proteus vulgaris* slant.

Procedure:

1. Disinfect the surface of working table and your hand with a disinfectant.
2. Take one of the broth culture and tap the bottom of the tube gently to prepare a suspension of the culture.
3. Hold the inoculation loop in your dominantly working hand and the broth culture in the other hand.
4. Sterilise the loop by holding the wire in the flame of Bunsen burner until it is red-hot. This is necessary to kill all the microorganisms present on the surface of inoculation loop.
5. While holding the loop like a pencil, grasp the cotton stopper by curling the little finger of the same hand. Do not place the cotton plug on the working bench, as it will cause contamination of broth culture.
6. Heat the mouth of the tube in the flame to kill the microorganisms present on the surface of tube around the lip.
7. Insert the sterilised, cooled inoculation loop into the broth culture to obtain a loopful of culture.
8. Remove the loop and flame the lip of the tube again before recapping. Keep the tube in test tube rack.
9. Take a tube of sterile nutrient broth and remove the cotton plug as described above. Flame the mouth of the tube and immerse the inoculating loop containing bacterial culture. Withdraw the loop, flame the mouth of tube and replace the cotton stopper. Keep the inoculated tube in the test tube rack.

10. Flame the loop until it is red hot and let it cool before placing on the working bench.
11. Label the inoculated tube with your name, date and name of bacterium inoculated.
12. Now take a nutrient agar slant and inoculate the slant moving the loop gently across the agar surface from the bottom of the slant to the top. Check that the loop does not penetrate and gouge the agar. Flame the mouth of the tube before putting the cotton plug. Flame your loop and cool it. Label the tube properly.
13. Inoculate the semisolid agar deep by inserting the inoculation needle straight down the middle of deep and pull it out gently. Do not forget to flame the lip of the tube and inoculation needle.
14. Inoculate a broth culture, agar slant and semisolid agar deep with other broth cultures one by one as described above.
15. Incubate all the tubes at 35°C for 48 hrs.
16. Record the appearance of each culture in nutrient broth, agar slant and semisolid agar deep.

Observations:

1. Describe the growth in nutrient broth as follows.

Bacterium	Turbidity	Presence of flocculant, pellicle or sediment	Pigment formation

2. Sketch the appearance of each culture on nutrient agar slant.
3. Locate the bacterial growth in semisolid nutrient agar deeps.

6.3. Serial Dilution Technique

In microbiological laboratories, dilution of microbial sample is done to isolate a bacterium in pure culture and to determine the number of microbes. Dilution need mixing a small amount of sample with a large volume of sterile water or normal saline also termed as **Dilution blank**. Since it is inconvenient to handle large volumes of sterile water, dilution is done in a series of smaller dilution blanks. This process is known as **Serial dilution**. A dilution is calculated as follows:

$$\text{Dilution} = \frac{\text{Volume of sample}}{\text{Total volume of sample and dilution blank}}$$

i.e. if 1 ml of sample is added into 9 ml dilution blank the dilution is represented as

$$\frac{1}{1+9} \text{ or } \frac{1}{10} \text{ or } 1:10 \text{ or } 10^{-1}$$

The total dilution in a serial dilution process is the product of all dilutions in a series. For example, if 1 ml is diluted with 9 ml and then 1 ml of this dilution is further put into a second 9 ml dilution blank, the final dilution will be

$$\frac{1}{10} \times \frac{1}{10} = \frac{1}{100} \text{ or } 1:100 \text{ or } 10^{-2}$$

Objective:

1. To learn to prepare dilutions of microorganisms.
2. To learn to interpret dilutions.

Materials required: Broth culture of *Escherichia coli*, Dilution blanks (9 & 99 ml), Sterile pipettes.

Procedure:

1. Pipette 1 ml of *E. coli* culture into a 9-ml dilution blank aseptically.
2. Mix the contents by rolling the tube between palms of both the hands.
3. Using a fresh pipette, transfer one ml of diluted culture into another 9-ml dilution blank.
4. Repeat the process till a desired dilution is obtained. The steps required to prepare 10^{-5} dilution using five 9-ml dilution blanks are depicted in Fig. 6.1a.
5. Remember that it is necessary to use a fresh pipette for each dilution in a series, however, any number of samples can be removed with one pipette from the same dilution blank.
6. Similarly, steps required to prepare 10^{-7} dilution by using three 99 ml dilution blank are depicted in Fig. 6.1b.

6.4. Isolation of Bacteria by Serial Dilution

Microorganisms are present in all types of natural environment (soil, water, food, air or animal/plant body) as mixed population. Such mixed cultures are of no use in studying microorganisms or their activities. To study cultural, morphological and physiological characteristics of an individual species, it is necessary to separate the organism under study from other species. Such a culture that contains a single kind of microbial species in known as **Pure culture**. Since bacteria are too small to separate directly without sophisticated micromanipulation equipment, indirect methods are used for separating them.

The two techniques generally used for the isolation of a pure culture of bacteria are **Streak plate** and **Pour plate**. Basic principle involved in both the methods is separation of microorganisms present in any environment by dilution and immobilizing them to enable them to grow and form a visible colony.

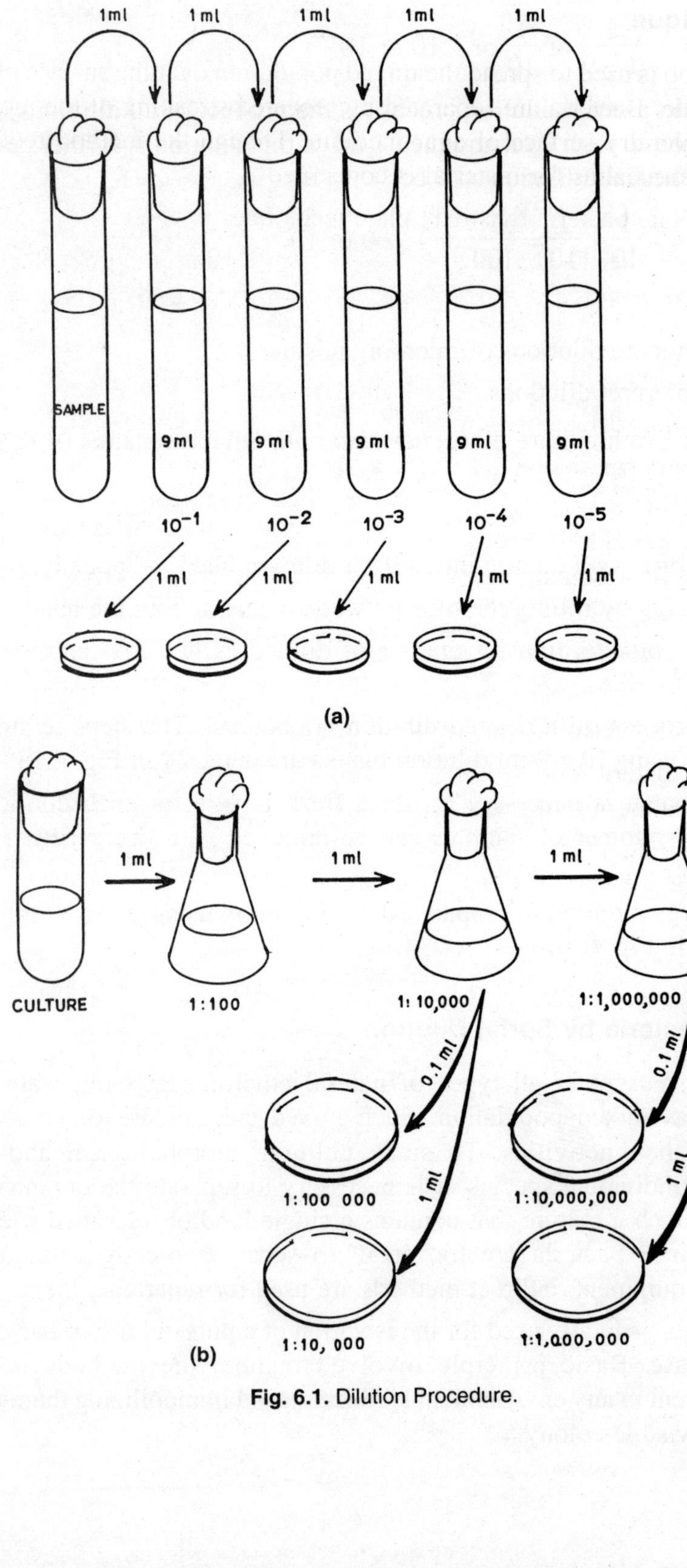

Fig. 6.1. Dilution Procedure.

Streak Plate Technique

In this technique, a loop is used to spread the mixed population over the surface of solid culture medium in a Petri plate. Bacteria are separated by repeated streaking of the inoculation loop and get immobilized on dry surface of agar medium. Though the method requires skill but save time and is economical as for material is concerned.

Objectives: To isolate bacteria by streak plate technique.

Materials required: Sterile Petri plates, Nutrient agar, Bunsen burner, Inoculation loop, Mixed culture of *Serratia marcescens*, *Micrococcus luteus* and *Escherichia coli.*

Procedure:

1. Disinfect the surface of the working table and your hands with a disinfectant.
2. Label the bottom surface of sterile Petri plates with your name and date.
3. Liquify nutrient agar, cool to 50°C and pour 20-25 ml medium into the bottoms of the Petri plates.
4. Be sure that you work near the flame of Bunsen burner in the aseptic zone and flame the neck of the flask containing nutrient agar prior to pouring to destroy microorganism present around the rim of the flask.
5. After pouring the medium into the plate, gently rotate the plate to distribute the medium evenly. Avoid splashing of medium up over the sides of Petri plate.
6. Allow the medium into the plates to solidify.
7. Hold the inoculation loop and heat in the flame of the burner to red-hot. Allow it to cool and take a loopful of mixed culture aseptically.
8. Streak the mixed culture on the solid surface of culture medium by lifting the cover of the Petri plate from one edge (Fig. 6.2).

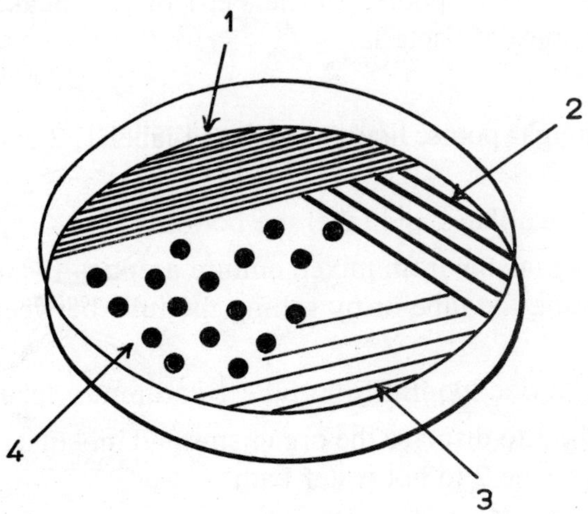

Fig. 6.2. Streak plate technique.

9. First streak over the area 1 near the edge of the plate.
10. Flame the loop and let it cool. Turn the plate slightly so that next sector 2 is on the top. Make several streaks from Area 1 through Area 2.
11. Flame the loop, cool it, turn the plate and streak from Area 2 through Area 3.
12. Repeat and streak as many times as possible from Area 3 through Area 4.
13. Flame the loop before keeping it down on the working table.
14. Incubate the plate in inverted position at 25°C for 24-48 hours.

Results: Well-isolated colonies of red, white and yellow colour will be observed in a nicely streaked plate.

Observations: Look for well-isolated colony and pick colonies of different colour and morphology. Prepare smear and perform Gram staining.

Precautions:

1. Apply as little pressure as possible on the inoculation loop to avoid gouging the agar while streaking.
2. Always flame the loop to redness before and after use.
3. Always work in aseptic zone, around the flame of Bunsen burner.

Pour Plate Technique

In this method a loop full of mixed culture is mixed with melted and cooled (45°C) culture medium containing agar and poured into empty, sterile Petri plates. After incubation, bacterial growth is visible as colonies in and on the agar of a pour plate. This method requires less skill compared to streak plate technique. However, it requires more media, tubes and plates.

Objectives: To prepare a pure culture of a bacterium by pour plate.

Materials required: Nutrient agar pours, Sterile Petri plates, Beaker of hot water (50°C), Inoculation loop, Mixed culture of bacteria.

Procedure:

1. Take three nutrient agar pours, liquify, cool and label 1, 2 and 3. Place them in a hot water bath at 50°C.
2. Take three sterile Petri plates and label the bottoms.
3. Take a loopful of organisms from mixed culture aseptically and inoculate tube 1. Mix thoroughly by shaking the tube or by rolling the tube between the plams of both the hands.
4. Inoculate tube 2 with one loopful from tube 1 and return tube 1 in hot water beaker.
5. Agitate and mix tube 2 to disperse the organisms and inoculate tube 3 with one loopful from tube 2. Return tube 2 to hot water bath.
6. Mix tube 3 thoroughly, flame its neck and pour the contents into a Petri plate. Rotate the Petri plate to spread the nutrient agar uniformly.

7. Flame the necks of tube 1 and 2 and pour their contents into their respective plates.
8. Allow the medium to completely solidify.
9. Incubate the plates at 25°C for 24-48 h in inverted position.

Results: Well-isolated colonies of different colour and morphology will appear in or on the surface of agar medium.

Observations: Select a plate with well-isolated colonies. Pick some colonies and perform Gram staining.

6.5. Enrichment Culture—Isolation of Photosynthetic Bacteria

Sometimes a desired microorganism is present in a sample in such a small number that it becomes difficult if not impossible to isolate it in pure culture. To isolate organisms found in minority, **Enrichment culture** technique is used whereby such environmental conditions (including medium) are provided that enhance the growth of desired bacteria. Other types of bacteria may also grow but their rate of growth will be slow. One such example is number of photosynthetic bacteria in soil.

The photosynthetic bacteria differ from algae and green plants in that these bacteria are anaerobic and utilize **Bacteriochlorophyll** instead of **Chlorophyll a** for photosynthesis. While the algae produce oxygen as a bye product of photosynthesis, the photosynthetic bacteria belonging to family **Chromatiaceae** and **Chlorobiaceae** produce sulphur. In bacterial photosynthesis H_2S acts as an electron donor and the chemical reaction proceed as follows:

$$CO_2 + 2H_2S \xrightarrow[\text{Sunlight}]{\text{Bacteriochlorophyll}} C(H_2O) + 2S + H_2O$$

The two families also differ in location of sulphur granules. **Chlorobium** produce extracellular sulphur granules and **Chromatium** produce intracellular sulphur granules. They also differ in pigmentation. The enrichment media for the two genera contain the same ingradients with difference being only in the amount.

Objectives:

1. To set up enrichment culture for enrichment of photosynthetic bacteria.
2. To isolate photosynthetic bacteria.

Materials required: Chromatium enrichment medium (non-sterile), Chlorobium enrichment medium (non-sterile), Pond muds, Table lamp (40-watt bulb).

Procedure:

1. Prepare and distribute each type of medium in 500 ml screw capped bottles.
2. Add enough pond mud into each type of medium so that pond mud forms a layer of 0.5-1 cm at the bottom.
3. Fill the bottles with medium to maximum capacity so that no air bubble is seen inside and tighten the screw caps.
4. Mix the mud with the media by inverting the bottles several times.

5. Place the bottles near to a table lamp at room temperature for seven days. The lamp should remain on 24 hours every day.
6. Examine the bottles each day and note for change of colour.
7. Prepare hanging drop slide of organisms from each bottle and examine using oil-immersion objective or a phase contrast microscope.

Results:

1. Change of colour of chromatium medium to red is an indication of presence of genus *Chromatium*.
2. Change of colour of chlorobium medium to green indicates the presence of genus *Chlorobium*.

6.6. Bacterial Population Count—Standard Plate Count

Number of bacteria present in water, milk or food material is an indicator of their handling, processing and ultimately quality. Thus it is sometimes useful to determine the number of organisms that are present in a unit volume. For this purpose several methods are available, which include **Direct microscopic count**, **Most probable number (MPN)**, **Standard plate count** and **Turbidity measurements**. Whereas direct microscopic count and turbidity measurement give total bacterial count including dead cells, the standard plate counts reveals actual viable number of organisms. Most probable number is useful to determine counts of gas forming bacteria in water, milk and food testing.

Standard Plate Count

Standard plate count is useful in determining the number of viable organisms in soil, water, milk, food or even in a bacterial or yeast culture. The procedure consists of diluting the organisms in a series of sterile dilution blanks. From blanks, measured amounts of the diluted organisms are transferred, spread over nutrient agar plate and incubated for 24-48 h. Thereafter plates are examined and number of colonies are counted to calculate the number of viable cells per unit volume of material under test.

Objectives:

1. To familiarize with the dilution technique and learn to prepare and interpret dilutions.
2. To determine the number of viable cells in a bacterial culture.

Materials required: Sterile dilution blanks (9 & 99 ml), Sterile 1 ml pipettes, Nutrient agar, Sterile Petri plates. *Escherichia coli* culture broth.

Procedure:

1. Liquefy nutrient agar.
2. Take three 99-ml sterile dilution blanks and label A, B and C.
3. Also label 8 Petri plates 1:10000, 1:100000, 1:1000000 and 1:10000000 in duplicate.

4. Aseptically transfer 1 ml of culture of *E. coli* into water blank A using 1 ml sterile pipette.
5. After shaking blank A vigorously, transfer one ml from blank A to blank B.
6. Shake blank B vigorously and transfer 0.1 ml to Petri plates labelled 1:100000 and 1.0 ml to the plates marked 1:10000 using a sterile 1-ml pipette. Also transfer 1.0 ml to blank C with the same pipette.
7. After shaking blank C transfer 0.1 ml to each plate marked 1:10000000 and 1.0 ml to each plate marked 1:1000000.
8. Cool down nutrient agar in a water bath maintained at 50°C.
9. Pour approximately 25 ml of nutrient agar into each of eight Petri plates. Rotate the plates gently to mix the organism in the medium.
10. After the nutrient agar solidifies, incubate at 25-30°C for 48 h in inverted position.

Results: Well isolated colonies of *E. coli* in and on the nutrient agar medium varying in number will appear after 48 h.

Observations:

1. Compare all the plates for number of colonies per plate and select plates that have colonies varying from 30-300 per plate.
2. Count all the colonies not caring for their size using a colony counter or magnifying lens.
3. Calculate the number of organism per ml of culture sample by multiplying the number of colonies counted by dilution factor and record your results.

For example

For example	Number of colonies per plate	= 185
	Dilution	= 1:100000
	Number of bacteria per ml	= 185 × 100000
		= 18500000
		= 18.5×10^6

6.7. Study of Cultural Characteristics

The cultural characteristics of an organism pertain to its macroscopic appearance on different kinds of media. The growth of an organism on a slant, a Petri plate and a broth is typical and varies from organism to organism and the media used. Most commonly used media for cultural studies are nutrient agar, nutrient broth and nutrient gelatin. Sometimes thioglycollate broth is inoculated to determine the oxygen requirements of an organism. Specific descriptive terms are used for describing the growth on different media.

Nutrient Agar Slant: The nutrient agar slants are inoculated in a single straight line and the growth of an organism is evaluated for the following criteria.

Growth: The amount of growth may be described as none, slight, moderate or abundant.

Pigmentation: Most of the organisms exhibit white or colourless growth whereas some depict shades of different colours. A few bacteria produce extracellular soluble pigments that diffuse into the medium and produce a colour.

Opacity: Degree of opacity may be expressed as opaque, transparent or translucent.

Form: The appearance of different types of growth on slants streaked in a single line is described as follows (Fig. 6.3).

Filiform	Uniform growth along the line of inoculation with smooth edges
Beaded	Distinct colonies along the line of inoculation
Arborescent	Branched, tree-like growth
Effuse	Growth thin and unusually spreading
Rhizoid	Root-like growth
Echinulate	Margins toothed or irregular

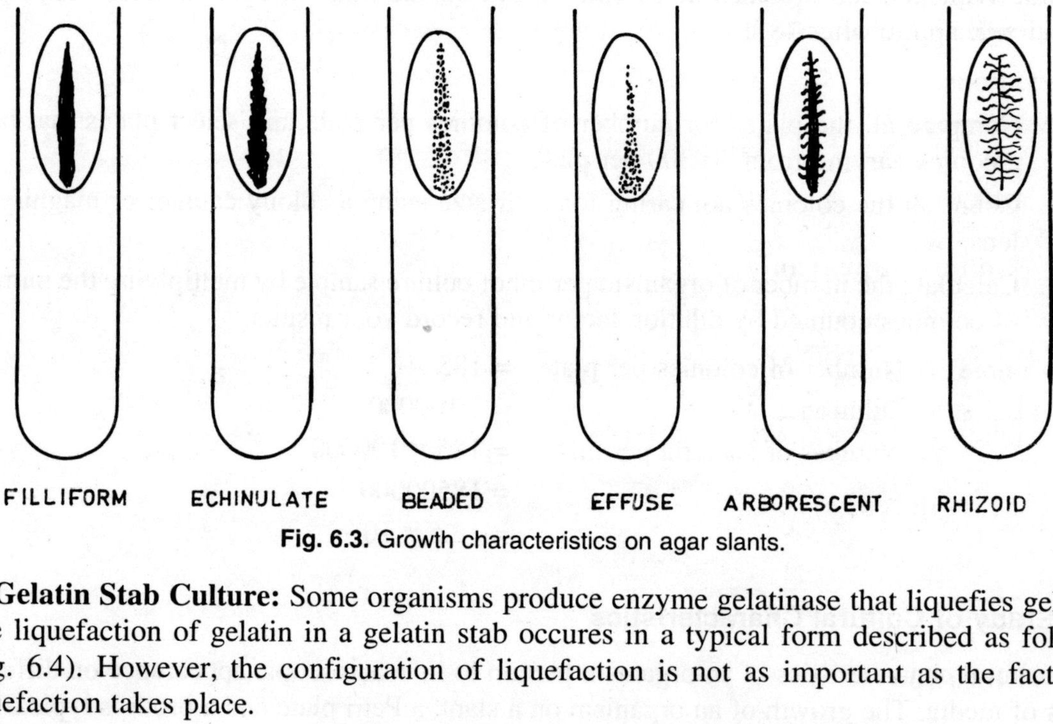

Fig. 6.3. Growth characteristics on agar slants.

Gelatin Stab Culture: Some organisms produce enzyme gelatinase that liquefies gelatin. The liquefaction of gelatin in a gelatin stab occures in a typical form described as follows (Fig. 6.4). However, the configuration of liquefaction is not as important as the fact that liquefaction takes place.

Crateriform	Liquefied surface area is saucer-shaped
Napiform	Liquefied surface area is turnip-like
Infundibuliform	Funnel-like
Saccate	Tubular, cylindrical
Stratiform	Complete liquefaction upto the walls of the tube

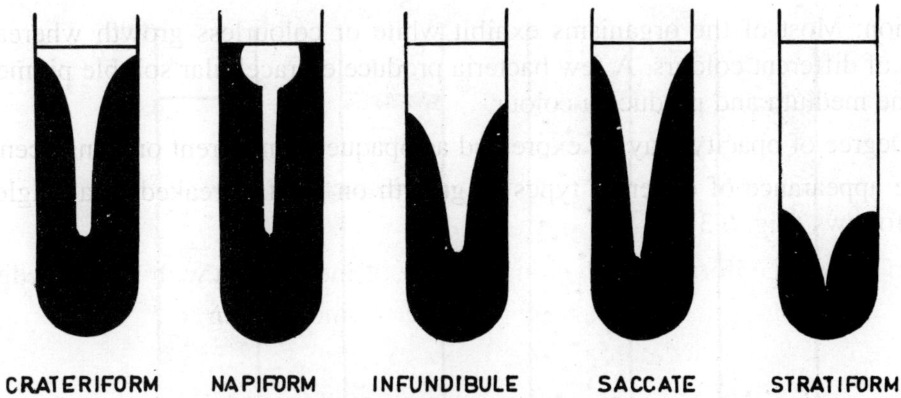

Fig. 6.4. Gelatin liquifaction.

Nutrient Agar Plate Culture: Colonies grown on plates of nutrient agar are studied and their size, colour, opacity, form, margin and elevation are observed. A hand lens may be used to study individual colonies. Some colony characteristics are given in Fig. 6.5.

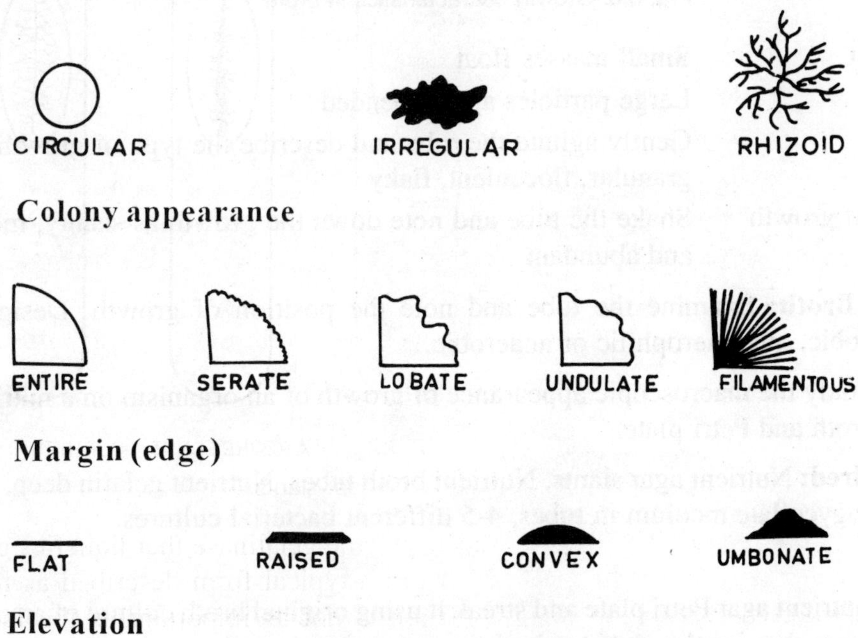

Fig. 6.5. Colony characteristics on solid surface.

Nutrient Broth: The nature of growth on surface, subsurface and bottom of the tube is observed and described as to the distribution and appearance of growth (Fig. 6.6).

Surface	Membranous, Ring, Pellicle, Flocculent
Subsurface	Turbid—when cloudy
Granular	Small particles are seen

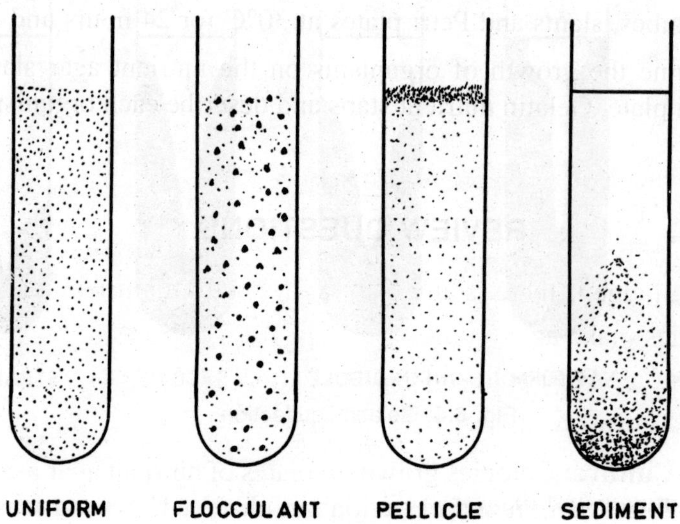

Fig. 6.6. Growth characteristics in broth.

Flocculent	Small masses float
Flaky	Large particles are suspended
Sediment	Gently agitate the tube and describe the type of growth as granular, flocculent, flaky
Amount of growth	Shake the tube and note down the growth as scanty, moderate and abundant

Thioglycollate Broth: Examine the tube and note the position of growth. Designate the organism as aerobic, microaerophilic or anaerobic.

Objective: To study the macroscopic appearance of growth of an organism on a nutrient agar slant, nutrient broth and Petri plate.

Materials required: Nutrient agar slants, Nutrient broth tubes, Nutrient gelatin deep, Nutrient agar plates, Thiogycollate medium in tubes, 4-5 different bacterial cultures.

Procedure:

1. Take a nutrient agar Petri plate and streak it using original broth culture of an organism in a manner that good isolated colonies appear after growth.
2. Streak nutrient agar slants with each of the culture in a straight line using inoculation needle.
3. Inoculate nutrient broth tubes using an inoculation loop.
4. Inoculate gelatin deep by stabbing the inoculation needle (straight) directly into the gelatin nutrient medium without disturbing the medium.
5. Inoculate tubes of thioglycollate broth with an inoculation needle. Mix thoroughly by rolling the tubes between palms of both the hands.

6. Incubate the tubes, slants and Petri plates at 30°C for 24 hours and examine.

Observations: Examine the growth of organisms on the nutrient agar slant, nutrient broth tubes, surface of Petri plate, Gelatin nutrient stabs and describe each organism using the terms mentioned above.

REVIEW QUESTIONS

1. How will you judge if there is any bacterial growth in nutrient medium?
2. What properties make agar a universal solidifying agent of culture media?
3. How important are Petri plates in microbiology?
4. Some bacteria grow on streak plate but do not when isolated by pour plate technique, Why?
5. Explain advantages and disadvantages of streak and pour plate techniques.
6. Why flaming of the inoculation needle is essential prior to and after each inoculation?
7. Why is it essential to hold the test tube cap/cotton plug in the hand during inoculation/transfer?
8. Why the inoculation needle is cooled prior to obtaining the inoculum?
9. Why is it essential to flame the neck of the tube/flask immediately after uncapping and before recapping?
10. Straight inoculation needle is used to inoculate agar deep tubes, why?
11. How composition of media affects bacterial growth?
12. Explain the principle underlying enrichment culture technique.
13. Differentiate between a complex and chemically defined/synthetic medium.
14. Why are complex media preferred over chemically defined media for routine cultivation of microorganisms?
15. Explain the selective advantages of each of the following medium: (a) Blood agar, (b) Eosin-methylene blue agar, (c) MacConkey agar, (d) Mannitol salt agar.
16. Explain the purpose of the following: (a) Blood in blood agar medium, (b) Lactose in MacConkey agar medium, (c) High salt concentration in Mannitol salt agar medium.
17. What are the advantages of serial dilution technique?
18. Why a fresh pipette should be used for each transfer in serial dilution technique?
19. How will you inhibit the growth of fungi for enumeration of bacteria in soil?
20. How will you control the growth of bacteria while isolating fungi/ yeast?
21. How will you proceed to isolate aerobic spore forming bacteria from soil?
22. In spread plate method of inoculation, the use of diluted cultures is advised, why?
23. Why the glass rod dipped in alcohol be held in downward position during flaming?
24. Describe how dilution of the culture is accomplished in streak plate method?

25. Distinguish between dilution and dilution factor.
26. Determine the number of microorganisms present in 1 ml of a suspension if 0.1 ml of a 1×10^{-5} dilution plate contains 126 colonies.
27. Write down the advantages and disadvantages of the serial dilution agar plate method.
28. Distinguish between selective and differential media.

Chapter 7

Physiological Characteristics

While studying morphological and cultural characteristics of bacteria, you will find that many bacteria share the same colony and cell morphology. Thus to identify and classify them further some physiological tests are performed. These tests have been designed on the basis of substrate a particular bacterium uses and the products formed. Organic molecules such as carbohydrates, proteins, peptides or lipids generally are the substrates, which are degraded by bacteria by enzymatic reactions either inside or outside the cell. All chemical reactions, which occur inside the cell, are termed as **Metabolism**. The chemical reactions that release energy and provide precursor molecules for the biosynthesis and assembly of macromolecules are known as **Catabolism** whereas chemical reactions involved in biosynthesis are referred to as **Anabolism**. Since all reactions are carried out by enzymes, the utilization of specific substrate and products formed are a measure of presence or absence of a specific enzyme system. These tests described in the following exercises are based on this principle.

Hydrolysis of Carbohydrates

Carbohydrates are grouped as Monosaccharides, Disaccharides and Polysaccharides on the basis of size of their molecules. Monosaccharides are simple sugars made of three to seven carbon atoms, disaccharides are composed of two molecules of monosaccharides and polysaccharides consist of eight or more monosaccharide units. While mono- and di-saccharides can enter the cell, however the polysaccharides generally cannot. Such high molecular weight substrates are first degraded to mono- or disaccharides by enzymes released by bacteria known as **Exoenzymes**. These small and low molecular weight mono- or disaccharides are transported inside and metabolised by enzymes, termed as **Endoenzymes**, present inside the cells. From simple carbohydrates like glucose bacteria derive energy by oxidation of glucose either by **Respiration** (aerobic or anaerobic) or by **Fermentation**.

Aerobic Respiration: A process where electrons released during oxidation of organic molecules are transferred to molecular oxygen via electron transport chain is called aerobic respiration. Electron transport chain consists of a series of cytochromes and ultimately electrons are passed on to molecular oxygen. Aerobic respiration is present in all aerobic organisms. Two commonly performed tests based on aerobic respiration are **Oxidase test** and **Catalase test**. There are four classes of bacterial cytochromes. In oxidase test the presence of Cytochrome c is determined. During aerobic respiration hydrogen peroxide is produced when hydrogen atoms combine with oxygen. Catalase is the enzyme that breaks down hydrogen peroxide to water and oxygen.

$$H_2 + O_2 \rightarrow H_2O$$

$$2H_2O_2 \xrightarrow{\text{Catalase}} 2H_2O + O_2$$

Anaerobic Respiration: In some organisms when molecular oxygen is not present, the electrons generated are transferred to inorganic compounds like nitrate, nitrite or nitrous oxide. Thus the **Nitrate reduction test** is an indicator of anaerobic respiration.

$$NO_3^- + 2H^+ + 2e^- \rightarrow NO_2^- + H_2O$$

$$2NO_2^- + 6H^+ + 6e^- \rightarrow N_2O + 3H_2O$$

$$N_2O + 2H^+ + 2e^- \rightarrow N_2 + H_2O$$

Fermentation: It is a process where donors as well as acceptor of electrons released during oxidation are organic molecules. The end products of fermentation are partially oxidised to small organic molecules like alcohols and organic acids. Although fermentation does not require oxygen but it can occur in the presence of oxygen. Since fermentation processes can produce a variety of end products depending upon the substrate, environmental conditions and the organism, many tests have been devised for the identification and differentiation of bacteria as follow:

Fermentation of different Sugars (Glucose, Sucrose, Lactose)

- Mixed Acid Fermentation (Methyl Red test)
- Butanediol Fermentation (Voges-Proskauer test)
- Hydrolysis of Proteins and Amino Acids

Proteins are organic molecules made of amino acids. Amino acids are joined together by peptide bonds to form a small chain (a peptide) or a larger molecule (a polypeptide or protein). The bacteria hydrolyse peptides and polypeptides with the help of exoenzymes and release amino acid which are transported inside and used as source of carbon and energy when carbohydrates are not available. The following tests are based on catabolism of proteins and amino acids.

- Casein Hydrolysis
- Gelatin Hydrolysis
- Deamination of Amino Acids
- Decarboxilation of Amino Acids

- Urease Test
- Indole Production
- Hydrogen Sulphide Production

7.1. Oxidase Test

Oxidase test is used to determine the presence of cytochrome c for differentiating certain groups of bacteria. For example, all the members of Enterobacteriaceae are oxidase-negative and most pseudomonads are oxidase-positive.

Objectives: To examine the presence of enzyme cytochrome oxidase.

Materials required: Trypticase soy agar plates, Oxidase test reagent, Cultures of *Pseudomonas aeruginosa* and *Escherichia coli.*

Procedure:

1. Divide one plate into two halves and lable one-half *E. coli* and the other *P. aeruginosa.*
2. Streak the two organisms on the appropriate half.
3. Incubate the plates at 35°C for 24 to 48 hours in inverted position.
4. To test for cytochrome oxidase, drop oxidase reagent on the colonies and observe for a change of colour.

Results: Colour of oxidase positive colonies will change to pink within one minute, then blue to black. No change in colour of colonies is indicative of oxidase negative.

Observations:

Organism	Colour after addition of reagent	Oxidase reaction (+/–)
Escherichia coli		
Pseudomonas aeruginosa		

7.2. Catalase Test

Catalase is essential for the breakdown of H_2O_2 produced during respiration. In anaerobes since catalase is absent, H_2O_2 inhibit their growth in the presence of oxygen. Oxygen gas released can be seen as white froth/effervescence if a few drops of 3% H_2O_2 are added to a drop of broth culture or a colony. All aerobic organisms give strong positive reaction. Some important catalase-negative bacteria are *Streptococcus*, *Leuconostoc*, *Lactobacillus* and *Clostridium.*

Objectives: To demonstrate the presence of catalse activity.

Materials required: Yeast extract slant, Yeast extract broth, Cultures of *Streptococcus faecalis* and *Staphyllococcus aureus.*

Procedure:

1. Inoculate yeast extract slants and broth with *Streptococcus faecalis* and *Staphyllococcus aureus.*
2. Incubate at 37°C.
3. Take a loopful of growth of each culture and keep on a slide.
4. Add a few drops of 3% H_2O_2 to the growth on slide and observe. Appearance of bubbles of oxygen is a positive reaction.
5. Add a few drops of 3% H_2O_2 to the broth culture and look for oxygen bubbles or froth accumulation on the surface.
6. In case of doubt, focus the slide with bacterial growth and 3% H_2O_2 under low power objective of a laboratory microscope and observe for emission of effervescence.

Results: Emission of effervescence because of production of oxygen from breakdown of H_2O_2 is a positive test.

Observations:

Organism	Appearance after addition of H_2O_2	Catalase reaction (+/–)
Streptococcus faecalis		
Staphyllococcus aureus		

7.3. Nitrate Reduction Test

Many facultative bacteria use inorganic nitrate as final electron acceptor in the absence of oxygen thus converting nitrate to nitrite. The reaction is catalysed by the enzyme known as **Nitrate-Reductase**. The bacterium under test is inoculated in nitrate broth and the nitrite formed is detected by the addition of dimethyl-α-naphthylamine and sulfanilic acid to nitrate broth. An appearance of red colour indicates the presence of nitrite.

Objectives: To identify facultative anaerobes by nitrate reduction test.

Materials required: Tubes containing nitrate broth, Nitrate reagent A (dimethyl-α-naphthylamine), Nitrate reagent-B (sulfanilic acid), Zinc dust, Cultures of *Lactococcus lactis*, *Pseudomonas aeruginosa* and *B. subtilis*.

Procedure:

1. Inoculate nitrate broth tubes with each of the three cultures in duplicate.
2. Incubate at 35°C for 48 hours.
3. After 48 hours add 5 drops of nitrate reagent A and 5 drops of nitrate reagent B to all the tubes inoculated with different cultures. Also add nitrate reagents A and B to an uninoculated control.

4. Shake the tubes gently. Appearance of red colour within 30 seconds is an indication of presence of nitrite.
5. To those tube which do not turn red add a pinch of zinc dust and record the appearance of red colour.

Results: A red colour indicating the presence of nitrite is a positive test. If reduction of nitrates has not occurred zinc dust reduces nitrate to nitrite. Appearance of red colour by addition of zinc dust is a nitrate reduction negative test.

Observations:

Organism	Colour after addition of nitrate reagents	Colour after addition of zinc	Nitrate reduction (+ or –)
Uninoculated control			
Lactococcus lactis			
Pseudomonas aeruginosa			
Bacillus subtilis			

7.4. Fermentation of Carbohydrates

Many bacteria ferment monosaccharides (glucose), disaccharides (sucrose, lactose) and polysaccharides (starch, cellulose etc.) and produce organic acids, hydrogen gas and carbon dioxide gas. A fermentation tube containing fermentation medium and desired carbohydrate is used to detect fermentation of carbohydrates. The indicator used is phenol red that is red at neutral pH and change to yellow colour in presence of acids. The gas produced is trapped in inverted Durham's tube.

Objectives: To perform and interpret fermentation of different carbohydrates.

Materials required: Glucose fermentation tubes, Lactose fermentation tubes, Sucrose fermentation tubes, 24 hour old cultures of *Escherichia coli*, *Aerobacter aerogenes*, *Proteus vulgaris* and *Bacillus subtilis*.

Procedure:

1. Inoculate two tubes of each medium with *E. coli*, *B. subtilis*, *P. vulgaris* and *A. aerogenes*. Also keep one tube of each medium as a control. Label each tube properly.
2. Incubate all the tubes at 35°C for 24-48 h.
3. Examine the cultures for the production of acid and gas and record the results.
4. Before recording a negative reaction, ensure that there is growth in the tube.

Results: Phenol red indicator gives red colour at neutral pH in an uninoculated fermentation tube. Fermentation that results in acid production will turn the indicator yellow (pH 6.8 or less). Gas if produced is trapped and appears as a bubble in Durham's tubes.

Observations:

Organism	Glucose		Lactose		Sucrose	
	Acid	Gas	Acid	Gas	Acid	Gas
Escherichia coli						
Bacillus subtilis						
Proteus vulgaris						
Aerobacter aerogenes						

7.5. Indole-Production

Indole a nitrogen-containing compound is formed by the degradation of amino acid tryptophan by certain bacteria and can easily be detected chemically. Thus it is a very good differentiation reaction among various bacteria. It is performed by inoculating a bacterium into tryptone broth and detecting indole by the addition of dimethylamino benzaldehyde (**Kovac's reagent**).

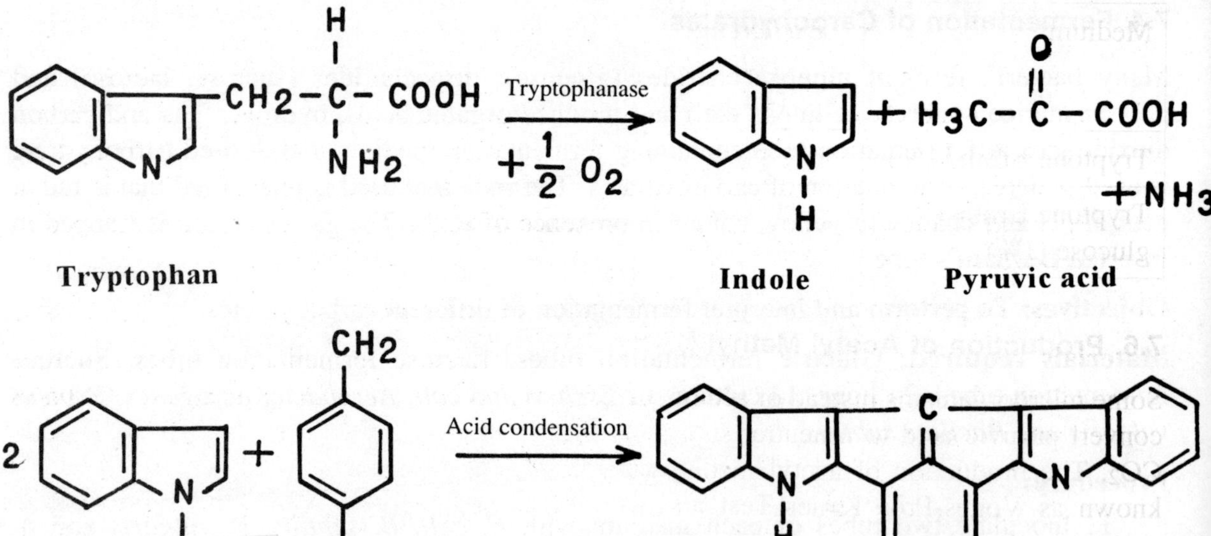

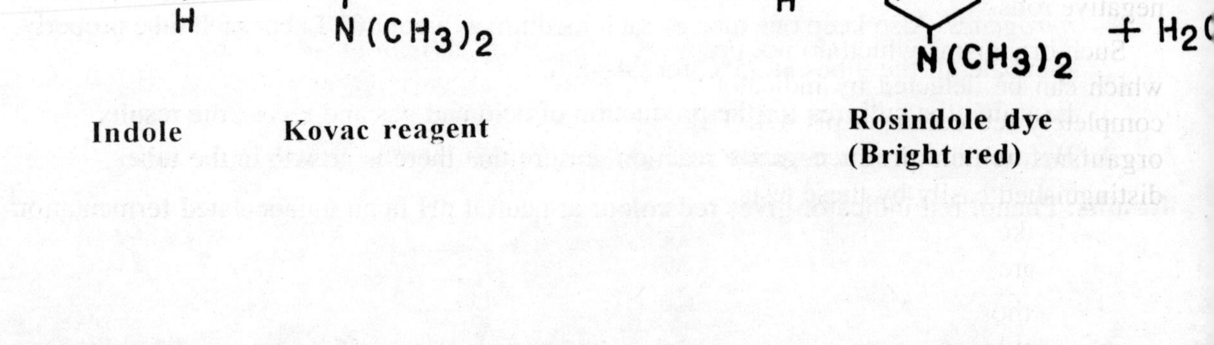

Objectives:

1. To demonstrate the formation of indole by degradation of proteins by bacteria.
2. To demonstrate that indole production may be used in differentiating various bacteria.

Materials required: Tryptone broth, Tryptone broth plus 1% glucose, Kovacs reagent, Cultures of *Escherichia coli*, *Bacillus subtilis*, *Aerobacter aerogenes* and *Streptococcus faecalis.*

Procedure:

1. Inoculate four tubes of tryptone broth with *E. coli*, *B. subtilis*, *A. aerogenes* and *S. faecalis.*
2. Inoculate a tube of 1% tryptone plus 1% glucose with *E. coli.*
3. Incubate at 37°C for 24-48 h.
4. Test for the presence of indole by adding approx. 1-2 ml of Kovac's solution. Shake gently. Allow to stand for some time.

Result: Reagent will rise to the top and the presence of indol is indicated by a cherry or deep red colour in the reagent layer. In the presence of carbohydrates, rapid production of acid inhibits the formation or activity of enzymes that degrade protein, thus lack of production of indole in presence of fermentable carbohydrates.

Observations:

Medium	*Escherchia coli* (+/–)	*Bacillus subtilis* (+/–)	*Aerobacter aerogenes* (+/–)	*Streptococcus faecolis* (+/–)
Tryptone Broth (1%)				
Tryptone Broth + glucose (1%)				

7.6. Production of Acetyl Methyl Carbinol and Methyl Red Test

Some microorganisms instead of accumulating acid products after the fermentation of glucose convert puruvic acid to a neutral substance acetylmethyl carbinol ($CH_3COCHOHCH_3$) and CO_2. The production of acetyl methyl carbinol can be detected by a simple test, which is known as Voges-Pros Kauer Test and is valuable in identifying *Bacillus* spp. and gram negative rods.

Such organisms which do not produce acetylmethyl carbinol lower the pH of the medium which can be detected by indicator methyl red which gives yellow colour at pH 6.0 and a completely red colour at pH 4.4. This is known as Methyl Red Test. Thus the two types of organisms which produce acids and neutral acetyl methylcarbinol (**Acetoin**) can be distinguished easily by these tests.

Objectives:

1. To differentiate closely related groups of bacteria.
2. To detect the formation of a specific intermediate product by fermentation of glucose by certain bacteria.

Materials required: Glucose peptone water, Barritt's Reagents, (Alpha naphthol solution and Potassium hydroxide solution), Methyl red indicator, Culture slants (24 h old) of *Escherichia coli, Aerobacter aerogenes.*

Procedure:

1. Inoculate two tubes of glucose peptone water with each of the culture. Keep at least one tube as control.
2. Incubate at 37°C for 48 h.
3. To determine the presence of acetyl methyl carbinol
   (a) Decant 1 ml of glucose peptone water into a clean test tube.
   (b) Add 0.5 ml of solution A (alpha-naphthol)
   (c) Add 0.5 ml of solution B (KOH)
   (d) Shake and allow to stand in the air for 5-15 min.
   (e) Repeat with the second culture and control
   (f) The appearance of pink red colour indicates the presence of acetyl methyl carbinol.
4. The production of acid may be detected by addition of 5 drops of methyl red indicator to the remaining glucose-peptone broth culture.

Results: Development of pink red colour after addition of Barrit's reagent is a positive test for acetyl methyl carbinol. The appearance of a pink red colour after addition of methyl red indicate a positive test for the production of acids and a yellow colour indicates the test negative.

Observations:

Organism	Acetylmethylcarbinol production		Methyl Red	
	Colour	+ or –	Colour	+ or –
Aerobacter aerogenes				
Escherichia coli				

7.7. Casein Hydrolysis

Casein is the prominent protein in milk. Some bacteria are capable of hydrolysing the milk protein which can be detected in litmus milk. Litmus milk that contains skim milk and litmus is opaque in appearance. The bacteria, which produce exoenzyme **Caseinase** hydrolyse casein to produce more soluble, transparent products like amino acids and peptides. This process of hydrolysis of casein is termed as **Peptonization** or **Proteolysis**. Further catabolism of amino acids results in alkaline reaction (purple colour) while production of acids by lactose fermentation reduce litmus turning litmus milk to white at the bottom of tube.

Obsectives: To perform and interpret different reactions in litmus milk.

Materials required: Tubes containing litmus milk, Cultures of *Pseudomonas aeruginosa* and *Proteus vulgaris*.

Procedure:

1. Inoculate two tubes with each of the culture.
2. Incubate the tubes of litmus milk at 35°C for 24-48 hours.
3. Note the change in colour of litmus milk.

Results:

Acid reaction	Pink colour of litmus milk indicates acid production and presence of fermentative bacteria.
Alkaline reaction	Blue or purple colour indicates presence of proteolytic bacteria.
Litmus reduction	Actively growing bacteria reduce oxidation/reduction potential and turn litmus milk white.
Peptonization	Proteolytic bacteria turn litmus milk translucent.

Observations:

Litmus milk	*Pseudomonas aeruginosa* (+/–)	*Proteus vulgaris* (+/–)
Acid		
Alkaline		
Reduction		
Peptonization		
No change		

7.8. Decarboxylation and Deamination of Amino Acids

Decarboxilation is the removal of carbon dioxide from an amino acid by the enzyme decarboxylase. The decarboxilation results in the formation of corresponding amine and a shift from a neutral to alkaline pH. The colour of the indicator Bromocresal purple changes from yellow to lavender purple in the presence of amines.

Once the amino acids are transported inside the cell, the amino group must be removed before it is used as a source of carbon and energy. The removal of amino group in the form

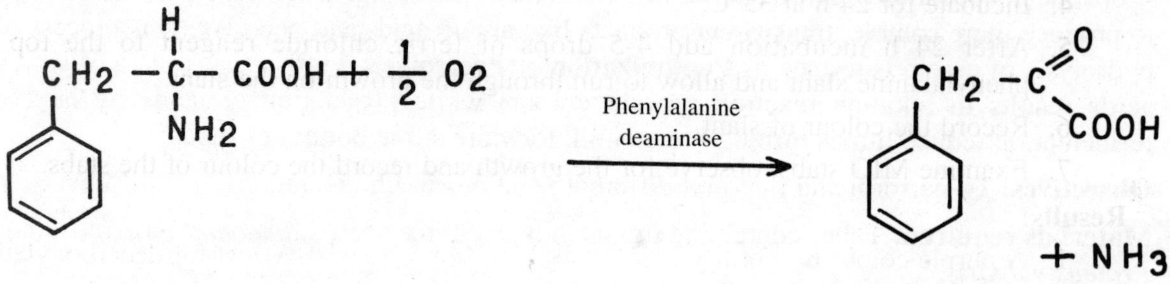

L-phenylalanine **Phenylpyruvic acid**

Phenylpyruvic acid + Ferric ion ⟶ Green complex

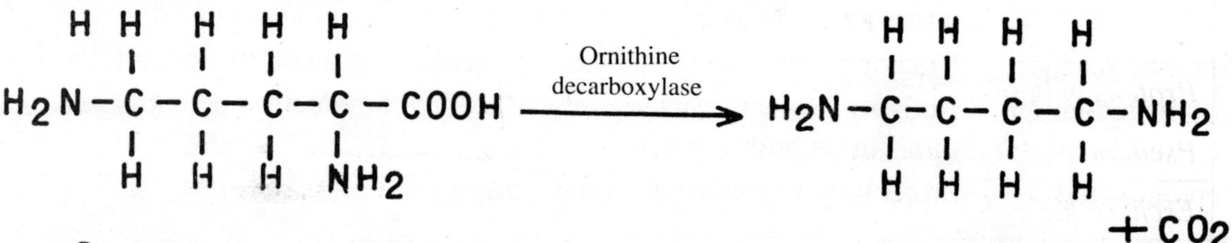

Ornithine **Putrescine**

Bromcresol purple (Yellow) **Bromcresol purple (Lavender-purple)**

of ammonia is called **Deamination**. Ammonia is excreted out of the cell and organic acid thus formed accumulates. Deamination of amino acid phenylalanine to phenylpyruvic acid can be detected by adding a solution of ferric chloride. Phenylpyruvic acid forms a complex with ferric ions which is green in colour. Deamination can also be detected with **Nessler's reagent** that gives deep yellow colour in presence of ammonia.

Objectives: To interpret the terms decarboxylation and deamination.

Materials required: Phenylalanine slants, Motility indole ornithine agar (MIO), Ferric chloride reagent, Cultures of *Escherichia coli*, *Proteus vulgaris*, *Pseudomonas aeruginosa* and *Enterobacter aerogenes*.

Procedure:

1. Take a phenylalanine slant and inoculate heavily with a culture of *Proteus vulgaris*.
2. Also streak one slant each with other cultures.
3. Stab one motility indole ornithine agar with each of the culture.

4. Incubate for 24 h at 35°C.
5. After 24 h incubation add 4-5 drops of ferric chloride reagent to the top of phenylalanine slant and allow to run through the growth on the slant.
6. Record the colour of slant.
7. Examine MIO stabs, observe for the growth and record the colour of the stabs.

Results:

1. A purple colour in motility indole ornithine stab is a positive test for decarboxylation while yellow colour indicates a negative test.
2. Appearance of dark green colour by addition of ferric chloride reagent in phenylalanine slant is a positive test for deamination.

Observations:

Organism	Phenylalanine deaminase			Ornithine decarboxilase		
	Growth	Colour	Reaction (+/–)	Growth	Colour	Reaction (+/–)
Proteus vulgaris						
Pseudomonas aeruginosa						
Escherichia coli						
Enterobacter aerogenes						

7.9. Hydrogen Sulfide Production

Degradation of sulfur containing amino acids frequently result in the liberation of H_2S. The production of H_2S by bacterial activities can be demonstrated in the laboratory if such cultures are grown on media containing salts of metals such as Lead, Bismuth and Iron. The darkening along the line of stab is caused by formation of metal sulfide.

Objective: To demonstrate the production of H_2S by bacteria grown on sulfur-containing proteins.

Materials required: Lead acetate agar stabs, Culture slants (24 h old) of *Bacillus subtilis*, *Proteus vulgaris* and *Escherichia coli.*

Procedure:

1. Prepare stab of lead acetate agar.
2. Inoculate with each of the culture in duplicate keeping a tube stabbed with sterile inoculation needle as control.
3. Incubate at 37°C for 48 h.
4. Examine the tubes after 48 hours for change of colour.

Results: Due to production of PbS the stabs get brown to black coloured indicating production of hydrogen sulfide.

Observations:

Organisms	H_2S production (+ or –)
Bacillus subtilis	
Proteus vulgaris	
Escherichia coli	

REVIEW QUESTIONS

1. What is the importance of production of indol? What is its source in the culture medium?
2. Is it possible to differentiate *E. coli* and *A. aerogenes* on the basis of indol test?
3. What are the products of degradation of proteins?
4. What is the practical importance of Voges-Proskauer test?
5. How acetylmethylcarbinol is produced from glucose?
6. How acid detected by methyl red test is produced?
7. What are the other products produced by *A. aerogenes* but not by *E. coli* from glucose in the medium?
8. Name some metal ions, which may be used, for detection of H_2S.
9. Name some microorganisms and conditions under which these microbes produce H_2S in nature.
10. List some gases, soluble or insoluble, which are released by micro-organisms.
11. Explain the temperature properties of gelatin.
12. Distinguish between respiration and fermentation.
13. Explain the purpose of Durham's tube in the fermentation vessel.
14. How microorganisms generate energy in the absence of carbohydrates?
15. Explain the purpose of phenol red in carbohydrate fermentation broth.
16. What are the metabolic end products of tryptophan degradation?
17. Describe the litmus milk reactions that may occur when proteins are metabolized as an energy source.
18. What are hydrolytic reactions?
19. Explain the functions of exoenzymes.

Chapter 8

The Bacterial Growth

Growth is an orderly increase in all the chemical constituents of an organism. Growth results in multiplication of cells, except in coenocytic organisms. It leads to increase in size of multicellular and increase in number of individuals in unicellular organisms. Growth may be measured either by determination of cell mass or cell number. A convenient method of measuring growth is determination of optical density.

The bacterial population growth studies require inoculation of viable cells into sterile broth medium and incubation of culture under optimum environmental conditions (pH, Temperature, O_2 etc.). Under these conditions, the cells reproduce rapidly and the dynamics of microbial growth can be plotted as increase in number versus time of incubation. Depending upon the availability of nutrient each individual microbe follows a typical pattern of growth. The curve thus obtained is known as growth curve. From the curve various stages of growth cycle can be ascertained.

Lag Phase: During this period there is no cell division and thus no increase in cell number. The cells adjust to the new environment. The cellular metabolism is accelerated and there is rapid biosynthesis of cellular macromolecules and primary enzymes. As a result the cell mass increases though slightly. The lag phase is more prominent when the inoculum used is either from stationary or death phase or the composition of the medium used is different from that of the inoculum.

Logarithmic (Log) Phase: Under optimum nutritional and environmental conditions, the cells reproduce at a uniform and rapid rate by binary fission. The increase in population is steep and exponential. The cells attain optimum size and the rate of growth is maximum during this phase. For all types of studies, cells must be used from logarithmic phase to get reproducible results. Growth rate of any organism is also determined from logarithmic phase.

Stationary Phase: During exponential phase of growth the nutrients are exhausted and toxic metabolites accumulate resulting in the decline followed by halt of growth. The number of cells undergoing division is equal to the number of cells dying. Thus there is no increase in cell number and population is maintained at its maximum level during this phase.

Death Phase: Soon the energy sources of the cells get depleted, metabolic wastes accumulate and the cells start dying at rapid but uniform rate. The decrease in population density is also exponential and parallels its increase during log phase. Theoretically, the entire population should die in a time equal to log phase. However, this never happens since a small number of highly resistant organisms always persist.

Mathematical Expression

The bacteria divide by binary fission and the growth in exponential phase can be expressed in a simple equation as follows.

For example, if we start from one cell (i.e. $N_0 = 1$)

Number of Divisions	(1)		(2)		(3)		(4)		(n)
Number of Cells	1	→	2	→	4	→	8	→	
	N_0		$N_0 \times 2$		$N_0 \times 2 \times 2$ or 2^2		$N_0 \times 2 \times 2 \times 2$ or 2^3		$N_0 \times 2^n$

Thus the number of cells (N_t) at time t in an exponential phase can be expressed as:

$$N_t = N_0 \times 2^n \text{ where n is the number of divisions.}$$

Logarithmic form of the above equation can be written as

$$\log N_t = \log N_0 + n \log 2$$

The number of divisions (n) is therefore:

$$n = \frac{\log N_t - \log N_0}{\log 2}$$

Growth rate constant (k) is defined as the number of divisions per unit time, thus:

$$k = \frac{n}{t} = \frac{\log N_t - \log N_0}{t \log 2}$$

Doubling or Mean Generation time (t_d) is the time required for a cell to divide and can be expressed as:

$$t_d \text{ or } g = \frac{t}{n} = \frac{1}{k}$$

The mean generation time can be determined either by direct method as described above or by indirect method (Fig. 8.1). To calculate the mean generation time by indirect method select any two points on the absorbance scale from the log phase such as 0.4 and 0.8 that represent a doubling of turbidity. Draw perpendicular lines from these points to their respective time interval on x-axis. From the information available determine mean generation time as follows:

$$g = t_{(O.\,D.\,0.8)} - t_{(O.\,D.\,0.4)}$$

$$g = 120 \text{ min.} - 80 \text{ min.} = 40 \text{ min.}$$

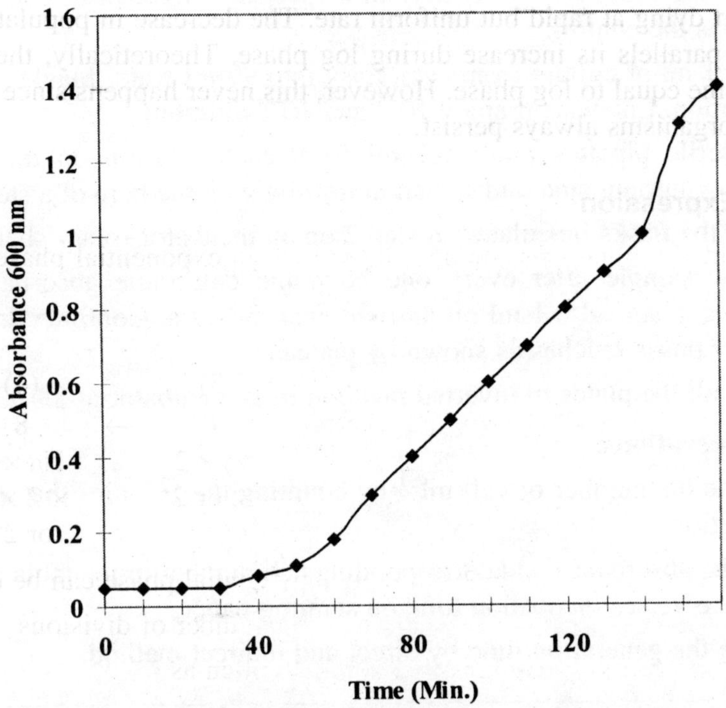

Fig, 8.1. Determination of generation time (Indirect method).

8.1. Growth Curve and Exponential Growth Rate Constant

Objectives:

1. To acquaint with population growth dynamics of bacterial culture.
2. To determine the growth rate constant and generation time of bacterial culture.

Materials required: Minimal medium broth culture of *Escherichia coli* (log phase), Minimal medium broth (200 ml in 500 ml Erlenmeyer's Flasks), Sterile water blanks (99 ml), Nutrient agar, Sterile Petri dishes, Sterile pipettes (1 & 10 ml.), Spectrophotometer, Cuvettes and Incubator rotary shaker.

Procedure:

1. Inoculate a loopful of *Escherichia coli* in 50 ml minimal broth and incubate on rotary shaker at 25-30°C for 24 hrs.
2. Next day, early in the morning inoculate 4 ml broth culture of *E. coli* (@ 2%) into each of the two 500 ml flasks containing 200 ml of minimal broth with the help of a sterile pipette.
3. Withdraw a sample of 2-4 ml from each flask using a sterile pipette (t_0). Take care that you flame the neck of flask and empty test tube properly to avoid contamination.

4. Measure absorbance with the help of a spectrophotometer at 600 nm using minimal medium as the control.
5. Transfer 1 ml of culture from each flask into 99-ml water blanks. This will give you 10^{-2} dilution. Likewise prepare 10^{-4} and 10^{-6} dilutions.
6. Using sterile pipettes place 0.1 ml from each dilution on the surface of plates containing nutrient agar and spread uniformly with the help of a sterile bent glass rod.
7. Incubate the flasks inoculated in step 2 on an incubator rotary shaker at 25-30°C.
8. Withdraw sample after every one hour and determine absorbance. Also prepare dilutions and spread 0.1 ml on nutrient agar medium. Continue drawing samples till stationary phase reaches as shown by plateau.
9. Incubate all the plates in inverted position in an incubator at 25-30°C for 24-48 hrs.

Results and observations:

1. Determine the number of cell ml^{-1} by counting the colonies that appear after 48 hrs. of incubation
2. Record the absorbance and corresponding cell number in the table and plot a graph of absorbance verses incubation time on semilog paper.
3. Calculate the generation time by direct and indirect method.

Incubation Time (Hrs.)	Absorbance (600 nm)	Number of Cells ml^{-1}	Log Number of Cells ml^{-1}
0			
1			
2			
3			
↓			
120			

8.2. Effect of Osmotic Pressure on Bacterial Growth

Osmotic pressure is the force that allows a solvent (water) to move from a solution of lower solute concentration to a solution of higher concentration across the semipermeable membrane. Microorganisms can live and grow in a **Hypotonic** environment in which the concentration of solute outside the cell is lower than inside the cell. A cell undergoes **Plasmolysis** i.e. water leaves the cell and membrane draws inward away from the cell wall, in an environment in which solute concentration outside the cell is higher than that inside the cell—a **Hypertonic** condition. A few bacteria are able to tolerate salt concentration upto 10% and are called **Facultative Halophiles**. Some others termed as **Extreme halophiles** require a minimum of 15-20% salt concentration for growth.

Since addition of solutes (salt or sugars) increases osmotic pressure, inhibit/halt growth or kill microorganisms, they are used to preserve foods and food products.

Objectives: To get acquainted with the effect of osmotic pressure on the growth of bacterial culture.

Materials required: Nutrient broth culture of *Escherichia coli* (24 hrs), Broth culture of *Halobacterium salinarium* (48 hrs), Nutrient agar containing 0.85%, 5%, 10%, and 25% sodium chloride, Sterile Petri dishes, Marking pencil.

Procedure:

1. Divide the underside of each of the five types of plates into two equal sections with a marking pencil and label each section with the name of the organism to be inoculated.
2. Inoculate each section of all the plates using a sterile inoculation needle with respective organism by making single line streaks.
3. Incubate all the plates in inverted position for 4-5 days at 25-30°C.

Observations and results: Observe each of the nutrient agar plate for the extent of growth of each culture and record as 0 = no growth, 1+ = scant growth, 2+ = moderate growth, 3+ = abundant growth.

Microbial Species	Sodium Chloride Concentration (%)				
	0.85	5	10	15	25
Escherichia coli					
Halobacterium salinarium					

8.3. Effect of pH on Bacterial growth

The acidity or alkalinity (pH) of the environment influences microbial growth. Most of the bacteria grow optimally between pH range of 6.5- 7.5. However, some of them can grow at an acidic pH below 4.0. Thus to preserve food organic acids are deliberately added. Some bacteria like lactic acid bacteria produce lactic and other organic acids and help in the preservation of food.

All microorganisms produce acids or alkaline substances and bring about changes in the pH in the surrounding environment, which sometimes prohibit the growth of microorganisms further. To neutralise the effect of acids or alkaline substances, buffering agents are added in the growth medium. Peptones present in the complex media act as buffers. In synthetic media combinations of Di-pottasium hydrogen phoshphate and Pottasium di-hydrogen phosphates are used to control the pH of the media.

Objectives: To explain the effect of pH on bacteria.

Material required: Nutrient broth adjusted to pH 2.5, 5.0, 7.0 and 9.5 in test tubes (15 × 100 mm), Actively growing cultures of *Staphyllococcus aureus*, *Escherichia coli*, *Serratia marcescens*, *Penicillium* or *Aspergillus* and *Saccharmyces cerevisiae.*

Procedure:

1. Inoculate a loopful of each organism into tubes of nutrient broth of different pH.
2. Incubate the tubes at 25-30°C for 24-48 hrs.

Observations and results: Observe all the tubes after incubation and record your results as presence (+) or absence (–) of growth of each organism.

pH	Microorganism				
	S. aureus	*E. coli*	*S. mercescens*	*Penicillium*	*S. cerevisiae*
2.5					
5.0					
7.0					
9.5					

8.4. Sporulation in Bacteria

Three different types of dormant structures, **Endospore** (*Bacillus, Clostridium, Sporosarcina*), **Cyst** (*Azotobacter*) and **Conidia** (actinomycetes) are formed in bacteria. The process of endospore formation in gram positive bacteria has been extensively examined. The endospore is a dormant structure that do not metabolise and differ from the vegetative cell in morphology, chemical composition and functions. A culture of *Bacillus* continue to grow vegetatively so long as nutrients are available and undergo morphological changes resulting in the formation of endospores under conditions when essential nutrients or water are not available. Endospore is surrounded by impervious layers called spore coats and is resistant to heat, chemicals, radiation, freezing, desiccation and other harsh environmental conditions. Sporogenesis in bacteria is merely a mechanism to overcome adverse environmental conditions and not a mean of reproduction.

During sporulation the forespore appears at one end of the cell and encloses a portion of the cytoplasm, a copy of the genetic material and eventually matures into a dormant spore distinct from the vegetative cell. Immediately before the onset of sporulation, the vegetative cells contain approximately twice the quantity of DNA as compared to that of the spore of the same species. Both the chromatin bodies are attached to mesosomes. The synthesis of DNA ceases at this stage, however, their shape and distribution continue to change until a single rope or **Axial** thread still attached to the mesosomes, within the cell is formed. The first definite sign of the beginning of sporulation is the formation of transverse septum near one pole of the cell which separates the cytoplasm and DNA of the smaller cell from rest of the cell contents. In the next step, the membrane of the larger cell rapidly grows around the smaller one and engulfs the smaller cell within the cytoplasm of the larger cell to produce a **Forespore**. Forespore is a protoplast enclosed by two unit membranes within the cell. At this stage the developmental process becomes irreversible and the cell is said to be '**Committed**' to undergo sporulation. Once the forespore has been formed, there is a rapid synthesis and development of **Cortex** between inner and outer membranes of the forespore. Shortly afterwards, spore coat appear exterior to the outer unit membrane surrounding the cortex. Although, spore coat has been synthesized and the spore has become refractile, it still is not heat resistant. Heat resistance develop only after massive uptake and deposition of Ca^{++} ions and dipicolinic acid, a compound not present in vegetative cells. Both are present in 1:1 ratio and represent 10-15% of spore dry weight. Dipicolinate start depositing after 16 hrs and the spore becomes heat resistant after 18-20 hrs. of onset of sporulation.

Objectives:

1. To examine the various cytological changes during sporogenesis.
2. To determine the development of heat resistance during endospore formation.

Materials required: Fresh culture of *Bacillus* spp. (Log phase), Nutrient agar, Sporulation medium (Tap Water 100 ml; Magnesium chloride, 10 mg; Calcium chloride, 20mg; Magnesium sulphate, 20 mg), Strile Petri plates, 99 ml Water blanks, Sterile test tubes, Sterile pippettes (1 & 10 ml), Alcohol, Glass spreader, Water bath, Thermometer, Glass slides, Microscope and Feulgen's stain.

Procedure:

Sporulation on solid medium:

1. Spread 1.0 ml of the fresh culture of *Bacillus* spp. on a nutrient agar plate in duplicate and incubate the plates at 30°C.
2. At intervals of 0, 0.5, 1, 2, 4, 6, 8, 10, 12, 24, 48, and 72 hours make smear on clean glass slides, heat fix and stain for chromatin material by Rainbow's method and draw diagrams.

Sporulation in liquid medium:

1. Inoculate 20 ml of 24 hrs old culture of *Bacillus* spp. into flasks containing sporulation medium.
2. Withdraw 1 ml samples in two sterile test tubes immediately after inoculation (0 hr. sample). Heat one of the sample to 90°C for 30 min.
3. Make serial dilutions (10^{-2}, 10^{-4} & 10^{-6}) from heated as well as control samples and spread 0.1 ml from each dilution on nutrient agar plates in duplicate. Incubate the plates at 30°C in inverted position.
4. Withdraw samples at 0.5, 1, 2, 4, 6, 8, 10, 12, 18, 24, 48 and 72 hrs. and repeat steps 2 and 3.
5. Make smears from each sample, stain by Rainbow's method for chromatin material and draw diagrams.

Results and observations:

1. Examine all the slides under a microscope using oil immersion objective lens and note down the cytological changes that occur. Draw diagrams.
2. Examine all the plates after incubation and determine the number of cell ml^{-1}. Record in the table and note the time for the development of heat resistance.

Time (hrs.)	Number of Cells ml^{-1}	
	Control	Heat-treated
0		
0.5		
1		
2		
4		
6		
↓		
48		
72		

REVIEW QUESTIONS

1. Define growth rate constant.
2. Why lag phase is observed during growth of bacteria? How can it be bypassed?
3. Endospores of *Bacillus* spp. are more resistant as compared to the spores of fungi, why?
4. How hypertonicity is useful in the control of microbial growth?
5. How do halophiles survive in their environment?
6. Is it possible to maintain a culture in logarithmic phase of growth indefinitely?
7. Why the pH of the medium changes during growth of microorganisms?
8. What are the methods used for the maintenance of pH of the growth medium?
9. Why bacteria undergo stationary phase?
10. Sporulation in bacteria is merely a mean of survival and is not a method of reproduction, comment?
11. How endospore of bacteria differs from a vegetative cell?
12. It is advisable to use inoculum from logarithmic phase of growth, why?
13. Define growth. How growth of unicellular organisms is measured?
14. What is meant by logarithmic growth? How logarithmic growth differs from arithmetic growth?
15. What is meant by generation time? How it is different from growth rate constant?
16. If a cell divides into four instead of two, how will you modify the following equation;

$$N_t = N_0 \times 2^n$$

17. What is plasmolysis?
18. How plasmolysis causes death of bacteria?

Chapter 9

Microbiology of Food

In ancient days people learnt from their experience that food should be cooked properly before eating and spoiled food should not be consumed. This probably was to protect the people from the danger of food borne diseases without any knowledge of their causative agent. This forced the people to adopt proper hygiene in handling the fresh meat and other food materials. Exact role of microorganisms in food spoilage and food borne diseases was known only after Pasteur's discovery of role of microorganisms in fermentation, putrefaction and disease. This led to the isolation of pathogens from food.

Food especially milk has served as primary vehicle for the transmission of gastrointestinal diseases. Foods get contaminated with soil microbes and microbiota from animals', food handlers and machinery during washing, grinding and packaging. Food being rich in organic nutrients is an excellent medium to support the growth and multiplication of microorganisms under suitable temperatures.

Food and dairy products may be contaminated in a variety of ways and from a variety of sources such as:

Soil and water: Food-borne microorganisms found in soil and water include members of the genera *Alcaligenes, Bacillus, Citrobacter, Proteus Clostridium, Pseudomonas, Serratia,, Enterobacter,* and *Micrococcus.* The common soil and water molds include *Rhizopus, Penicillium, Botrytis, Fusarium, and Trichothecium.*

Enteric microorganisms of humans and animals: The major members of this group are *Bacteroides, Lactobacillus, Clostridium, Escherichia, Salmonella, Proteus, Shigella, Stphylococcus,* and *Streptococcus.*

Food handlers: People who handle foods are likely to contaminate them because microorganisms present on their hands and clothing are easily transmitted. A major offending organism is *Staphylococcus,* which is generally found on hands, on skin, and in the upper respiratory tract. Food handlers with poor personal hygiene and unsanitary habits are most likely to contaminate foods with enteric organisms.

Utensils: Utensils used for cooking and storage of food should be washed with an effective detergent and clean hot water to prevent contamination of food. The type of microorganisms depends on the type of food and the manner in which the utensils were handled.

Animal hides and feeds: Animal hides get contaminated with microorganisms present in water, soil, feed, dust, and fecal debris and these may serve as a potential source of infection for workers.

Today a large number of food borne diseases are known and their causative agents, methods of prevention and cure are well understood. On the basis of mode of illness, food borne diseases are classified into three groups as follows:

A. Intoxication: Illness or food poisoning occurs as a consequence of ingestion of food containing preformed bacterial or mold toxin. Several bacteria and molds produce toxins during growth in a food. Once a toxin has been produced in a food, there is no need of viable cells and the food containing toxin result in the illness e.g. Botulism, *Staphylococcus* food poisoning.

B. Infection: Illness occurs as a result of the consumption of food and water contaminated with enteropathogenic bacteria. It is necessary for the cells of enteropathogenic bacteria to remain alive in the food and water during consumption. The viable cells, even if present in small numbers, have the potential to establish and multiply in the digestive tract of human to cause illness e.g. Salmonellosis.

C. Toxicoinfection: Illness occurs from the ingestion of a large number of viable cells of some pathogenic bacteria through contaminated food and water. Generally the cells either sporulate or die and release toxin(s) to produce the symptoms e.g. *Clostridium perfrigens* gastroenteritis.

By enumerating microorganisms in milk and food, the quality of a particular sample can be determined. Although the microorganisms cannot be identified, the presence of a high number suggests a good possibility that pathogens may be present. Even if a sample contains a low microbial count, it can still transmit infection. Thus it becomes important to determine the microbiological quality of food during processing, storage and marketing.

9.1. Microbiological Analysis of Food Products: Bacterial Count

Microorganisms present in food are harmful in some cases, while in others they are definitely beneficial. Certain microorganisms are necessary in preparation of foods such as cheese, pickles, sauerkraut, yogurt, and sausage. The presence of harmful microorganisms is often responsible for serious and sometimes fatal food poisoning and toxicity or spoilage of food. In milk or water, the presence and number of coliform bacteria and other enteric organisms is indicative of fecal contamination and presence of pathogens.

Objectives:

1. To determine the viable cell number of microorganisms present in food products.
2. To determine the presence of coliform bacteria in the selected food products.

Materials required: Samples of thawed frozen vegetables, dried fruit or any other food product, Brain-heart infusion agar deep tubes, Eosin-methylene blue agar plates, Sterile water blanks (99 & 180 ml). Water bath, Balance, Sterile butter paper, Waring blender with three sterile jars, Sterile Petri dishes, Sterile pipettes (1ml), Inoculation loop, and Glassware marking pencil.

Procedure:

1. Label six Petri dishes for each of the food samples to be tested and their dilutions (10^{-2}, 10^{-3}, 10^{-4}) in duplicate. Label three eosin-methylene blue agar plates with the type of the food to be tested.
2. Melt the brain-heart infusion agar deep tubes, cool and maintain in a water bath at 45°C.
3. Weigh 20 g of each food sample on sterile butter paper and put into sterile waring blender jar. Add 180 ml of sterile water into the blender jar and blend the mixture for 5 minutes. In this manner prepare a 1:10 (10^{-1}) dilution of each food sample.
4. Transfer 1 ml of the 10^{-1} of ground food suspension into 99 ml sterile water blank, to prepare a 10^{-3} dilution, and add 0.1 ml to the appropriately labeled Petri dish.
5. Shake the 10^{-3}-sample dilution, and using a different pipette, transfer 1 ml to the plate labeled 10^{-3} and 0.1 ml to the plate labeled 10^{-4}.
6. Add a 15-ml aliquot of the molten and cooled agar to each of the three plates. Swirl the plates gently to obtain a uniform distribution and allow the plates to solidify.
7. Repeat steps 3 to 6 for the other food samples.
8. With the help of a sterile wireloop streak from 10^{-1} dilution of each food sample on its appropriately labeled eosin-methylene blue agar plate.
9. Incubate all the plates in an inverted position for 24 to 48 hours at 37°C.

Observations:

1. Count the number of colonies on valid plates that contain between 30 and 300 colonies.

Food Sample	Dilution	Number of Colonies (ml^{-1})	Number of organisms (ml^{-1})
A	10^{-2}		
	10^{-3}		
	10^{-4}		
B	10^{-2}		
	10^{-3}		
	10^{-4}		
C	10^{-2}		
	10^{-3}		
	10^{-4}		

2. Determine the number of organisms per milliliter of each food sample by multiplying the number of colonies counted by the dilution factor.

3. Record the number of colonies per plate and the number of microorganisms per milliliter of each food sample.

9.2. Standard Plate Count of Milk

Milk is an excellent medium for the growth of all types of bacteria if appropriate temperature is provided. The number of bacteria present in milk is a reliable indication of its sanitary quality. Therefore, in routine examination of bacteriological quality of milk, standard plate count is most appropriate technique.

Objectives:

1. To determine the viable number of bacterial cells in milk sample.
2. To determine the microbiological quality of milk.

Materials required: Milk sample, Sterile 99 ml dilution blanks, Sterile Petri plates, Sterile 1 ml pipettes, Tryptone glucose yeast extract agar.

Procedure:

1. Shake the milk sample well and prepare 1:100, 1:10000 and 1:1000000 dilutions as per method described in section 6.3.
2. Label 2 Petri plates each of the dilution 1:100, 1:1000, 1:10000, 1:100000, 1:1000000 and 1:10000000.
3. Add samples of diluted milk in appropriately marked plates.
4. Incubate the plates at 35°C for 24 hours and select plates having 30-300 colonies for counting.
5. Record your results and calculate the number of viable cells per ml of milk.

Results and observations:

Sample No. and source	Average No. of colonies per plate	Dilution	Number of organisms per ml
1			
2			
3			
4			

9.3. Direct Microscopic Count of Microorganisms in Milk

The number of microorganisms present in milk can also be determined directly by spreading a measured amount of milk over one square centimeter area on a Breed slide. The slide is examined under oil immersion objective after staining and number of microorganisms present in an entire microscopic field is counted. The method has certain advantages as compared to standard plate count. First of all, it is fast and quality of milk is determined in a much shorter time. Secondly, it will reveal the presence of bacteria that do not form colonies on agar medium at 35°C. Thirdly, it will also determine the number of leukocytes indicating that the milk is from diseased animal.

Since it has several advantages over the other methods, it is widely used for testing of raw milk at receiving centres. However, it requires a technically qualified person.

Objectives: To examine the bacteriological quality of raw milk by direct microscopic count.

Materials required: Breed slide, Methylene blue, Xylol, Ethyl alcohol (95%), Beaker of water, Hot plate, Sample of raw milk.

Procedure:

1. Calibrate a microscope using a stage micrometer and oil immersion objective as described in Section 3.3.
2. Shake the milk sample well to break up clumps of bacteria, if any.
3. Transfer 0.01 ml of milk sample to a Bread slide with a micropipette and spread over an area of one square centimeter marked on the slide.
4. Air dry the smear and place it over a beaker of boiling water for 5 minutes to fix the smear.
5. Cover the slide with xylol to remove fat globules.
6. Rinse with 95 per cent ethyl alcohol tc remove xylol.
7. Immerse the slide into a beaker containing distilled water to remove alcohol. Be careful that washing of slide under running tap water may wash off the milk film.
8. Stain the smear with methylene blue for 15 seconds. Remove excess of methylene blue by immersing the slide in distilled water.
9. Decolourize the smear to pale blue with 95 per cent alcohol and dip in istilled water to stop decolourization.
10. Air dry the slide, focus using oil immersion objective. Count the number of bacteria in several microscopic fields and determine the number of microorganismis per ml of milk as describe in section 3.3.

Results and observations:

Field diameter .... millimeters

Field Area .... square millimeter

No. of fields in one square centimeter ....

Microscopic factor (100 × No. of fields) ....

Milk Sample Number	Average No. of bacteria per field	Average No. of leucocytes per field	No. of bacteria per ml	No. of leucocytes per ml
1				
2				
3				
4				
5				

9.4. Microbiological Analysis of Milk: Reductase Test

Methylene blue is reduced and loses its colour in presence of actively growing bacteria. The time taken for the reduction of methylene blue is inversly proportional to the number of viable bacteria. Thus shorter the **Methylene blue reduction time** (MBRT), poorer is the quality of milk.

The test is based on the principle that all bacteria in milk lower the oxidation-reduction potential at 35°C. In raw milk bacteria like *Streptococcus lactis* and *Escherichia coli* are generally present. These bacteria are strong reducers and thus the test is suitable for examining milk at receiving centres. The advantage of the test is that a person with little technical training can also perform the test and determine the sanitary condition of the milk.

Objectives:

1. To get familiarize with the testing procedure of milk quality by reductase test.
2. To interpret the principle of reductase test.

Materials required: Raw milk samples, Sterile test tubes with stopper, Methylene blue (1:25000), Water bath (35°C), Sterile 10 ml pipettes, Sterile 1 ml pipettes.

Procedure:

1. Label sterile test tubes with your name and type of milk sample.
2. Transfer 10 ml of each milk sample into previously labelled test tubes using separate pipette for each sample.
3. Add 1-ml methylene blue with a 1-ml pipette into each of the milk tube. Insert rubber stopper and mix the contents by inverting the tubes several times.
4. Record the time on the tubes and place all the tubes in the water bath set at 35°C.
5. After five minutes of incubation, remove the tubes from water bath and invert once to mix.
6. Remove the test tubes from the water bath after every 30 minutes and note the change of colour of milk from blue to white.

Results: The quality of milk is expressed as follows:

Excellent	Methylene blue not decolourized in 8 hours.
Good	Methylene blue decolourized in less than 8 hours but greater than 6 hours.
Fair	Methylene blue decolourized in less than six hours but not less than 2 hours.
Poor	Methylene blue decolourized in less than 2 hours.
Very poor	Methylene blue decolourized in 30 minutes.

Observation: Record the time of reduction of methylene blue in each sample of milk.

9.5. Production of Wines From Grapes

Wine has been a part of the food of man ever since the settlement in the Tigris-Euphrates basin. It has also been used as medicine for centuries all over the world. It is a fermented alcoholic beverage produced from grapes and other fruits such as peaches, pears, plums, and apples by the action of yeast cells. This is a biochemical conversion of sugars (Glucose & Fructose) present in juice. Enzymes present in yeast degrade fruit sugars first to acetaldehyde and then to alcohol.

Grapes containing 20% to 30% sugar will yield wines with an alcohol content of approximately 10% to 15%. In addition, wines contain vitamins, amino acids, esters, sugars, minerals and tartaric acid. A large number of other chemical compounds derived from metabolic activities of yeast are also present in wines. Wines are of two types **Red** and **White** wine. For red wine, the crushed grape must is fermented with their skin to extract their colour into the juice. White wine is produced from the juice of white grapes.

The commercial production of wine is a long and exacting process. First, the grapes are crushed or pressed to extract the juice, which is called must. Potassium metabisulfite is added (@ 147 mg kg^{-1} grapes or 294 mg l^{-1} of grape juice equivalent to 100 ppm sulphur dioxide) to the must to retard the growth of acetic acid bacteria, molds, and wild yeast that are generally present on the surface of grapes. A wine-producing strain of yeast, *Saccharomyces cerevisiae* var. *ellipsoideus*, is inoculated into the must, which is then incubated for 3 to 5 days under aerobic conditions at 21°C to 32°C. This is followed by an anaerobic incubation period. The wine is then aged for a period of 1 to 5 years in aging tanks or wooden barrels. During this time, the wine is clarified of any turbidity, thereby producing volatile esters that are responsible for characteristic flavours. The clarified product is then filtered, pasteurized at 60°C for 30 minutes and bottled.

Total acidity: Total acidity is expressed as % tartaric acid. To a 10-ml aliquot of the fermenting wine, add 10 ml of distilled water and 5 drops of 1% phenolphthalein solution. Mix and titrate with 0.1 N sodium hydroxide till the first appearance of pink color.

Total acidity is calculated using the following formula:

$$\% \text{ Tartaric acid} = \frac{\text{Amount of NaOH} \times \text{Normality of NaOH} \times 7.5}{\text{Weight of sample in g}}$$

where 1 ml = 1 g

Volatile acidity (expressed as % acetic acid): Calculate volatile acidity using the following formula:

$$\% \text{ acetic acid} = \frac{\text{Amount of NaOH} \times \text{Normality of NaOH} \times 6.0}{\text{Weight of sample in g}}$$

where 1 ml = 1 g

Alcohol (expressed as volume %): Alcohol is estimated colorimetrically by the method of Caputi *et al.* (1968).

Reagents: Potassium dichromate solution: Dissolve 34 g potassium dichromate in 500 ml distilled water. Add 325 ml concentrated sulphuric acid slowly and make volume to 1000 ml with distilled water.

Procedure: Dilute 5 ml of sample to 50 ml with distilled water. Take 10 ml of the diluted sample in a distillation flask containing 25 ml distilled water and distill. Collect approximately 20 ml distillate in a volumetric flask containing 25 ml of potassium dichromate solution. Incubate the flasks in a water bath at 60°C for 20 min. to develop and stabilize the colour. Cool down to room temperature and make the volume by adding distilled water. Measure the absorbance of the sample at 600 nm using a spectrophotometer. The amount of ethanol present in the sample can be calculated by referring to a standard curve.

Aroma: Fruity, yeast-like, sweet, none.

Clarity: Clear, turbid.

Objectives: To acquaint the students with the process of wine production from grapes by fermentation.

Materials required: White grape juice broth culture of *Saccharomyces cerevisiae* var. *ellipsoideus* (50 ml) grown for 48 hours at 25°C, Pasteurized white grape juice (500 ml) or 1 kg fresh grapes, Phenolphthalein solution (1%), Sodium hydroxide (0.1 N), Sucrose, Erlenmeyer flask, Cotton, Pan balance, Spatula, Butter paper, Graduated cylinder (10 ml), and Burette or pipette for titration.

Procedure:

1. Take 1 kg white grapes, wash, remove the berries and crush to extract the juice. Sieve through a muslin cloth to separate the pulp.
2. Pour 500 ml of the white grape juice into a 1 liter Erlenmeyer flask and add 20 g sucrose.
3. Inoculate 50 ml of *S. cerevisiae* culture grown in grape juice broth (10% starter culture). Plug the flask with cotton.
4. After 2 and 4 days of incubation, add 20 g sucrose to the fermenting wine.

5. Incubate the fermenting wine for 15 days or until bubbling stops at 25°C.

Observations:

1. Determine total acidity and volatile acidity in uninoculated grape juice and fermenting wine at an interval of 5 days.
2. Note aroma and clarity.
3. Determine % alcohol (optional).

Parameter	Grape Juice	Fermenting Wine		
		5 Days	10 Days	15 Days
Tartaric Acid (%)				
Acetic Acid (%)				
Alcohol (%)				
Aroma				
Clarity				

REVIEW QUESTIONS

1. Why the milk is spoiled if not stored at refrigeration temperature?
2. Define Pasteurisation. How does it differ from sterilisation?
3. If nutrient medium for the growth of a bacterium is pasteurised, what will happen?
4. Why milk get sour, though slowly, when stored in a refrigerator?
5. What is the function of methylene blue in reductase test?
6. How fermentation is useful in the preservation of food?
7. Name some industrially important fermentations.
8. Why personnel hygiene is important in food industry?
9. What is the importance of reductase test in dairy industry?
10. Discuss the advantages of Breed's counting technique.
11. Milk is considered to be a sterile body fluid, how it gets contaminated during and after milking?
12. What are the various possible ways in which food get contaminated with enteric organisms?
13. Does the presence of *Escherichia coli* in food indicative of contamination?
14. Why potassium metasulphite is added in the must for preparation of wine?
15. What are the chemical end products of fermentation of grapes?

16. How the preparation of red wine differs from that of white wine?
17. Give major difference in red and white wine.
18. Why is wine Pasteurised and not sterilised?
19. What occurs during aging of wines?
20. Why wines get sour?
21. Why Pasteurisation of milk is preferred over sterilisation?
22. Use of salt is an ancient process of food preservation, explain the mechanism by which the salt help in preventing the spoilage of food.

Chapter 10

Microorganisms and Environment

Microorganisms exist everywhere. It is easier to list the environments devoid of microorganisms but it is almost impossible to list environments containing one or another type of microorganism. They are present in the water we use for drinking and bathing, in our food, on the utensils, on our clothes and bed sheets. A wide variety of microorganisms are present in the air we breathe. They have been found in the air miles above the earth. Microbes are present in many unlikely and extreme environments because of their varied metabolic properties. They have been isolated from arctic region, they can thrive in natural hot water springs at temperature of more than 90°C. While some bacteria can grow in concentrated sulphuric acid others can survive and multiply in distilled or deionized water. Thus the microbes are widely distributed and there is not much difference in genera and species because microorganisms are easily carried and transported by air currents because of their negligible mass. Under natural conditions, no substrate or location need any inoculation (exception - materials sterilised and protected from air).

However, a majority of these microorganisms are beneficial to human beings. The most important role that is played by microorganisms is in bio-geochemical transformations. As decomposers microbes are responsible for recycling of nutrients. Simple inorganic nutrients are assimilated into complex organic compounds by photosynthetic organisms, which serve as the ultimate food source for all consumers. Microorganisms recycle these nutrients trapped in the bodies of dead animals and plants by mineralisation. This recycling allow the sustainance of life on the earth with limited supply of nutrients. Thus the microorganisms are responsible for sustainance of life on the earth planet.

10.1. Microbes in the Environment—Ubiquitous Presence

Microbes are found in water we drink the air we breathe and the food we eat. They live in and on our bodies. Thus microbes are present everywhere and to maintain aseptic conditions while working in a microbiological laboratory, one must keep this fact in mind. In this exercise you will examine the widespread distribution of bacteria in the environment.

Objectives:

1. To demonstrate the presence of microorganisms in and around the laboratory.
2. To demonstrate the presence of microorganisms on some of our body parts.

Materials required: Petri plates containing nutrient agar, Tubes containing nutrient broth, Sterile cotton swabs, Sterile water.

Procedure:

1. Take several plates of nutrient agar and expose in the following manner.
   (a) Leave the nutrient agar plate open to air for 30 minutes in the laboratory.
   (b) Keep the nutrient agar plate open to air for 30 minutes outside the building.
   (c) Blow dust onto exposed nutrient agar.
   (d) Press coins on the nutrient agar for a short period.
   (e) Place a hair on the agar.
2. Moisten a sterile cotton swab with sterile water. First swab the surface of your working bench or floor or any object and then the surface of nutrient agar with the same swab. Inoculate several plates from different environment.
3. After swabbing the nutrient agar surfaces transfer the swab into nutrient broth and leave it there.
4. Incubate the plates and tubes at 35°C for 24-48 hours in a incubator in inverted position.

Results: Examine the nutrient agar plates and nutrient broth tubes for the growth of microorganisms. You will observe colonies of bacteria on the agar surface. Cloudiness in nutrient broth after shaking compared to uninoculated tube of broth is positive for growth.

Observations:

1. Count the number of colonies of bacteria and fungi on the plates separately.
2. After shaking the tube of broth, look for turbidity and record your observations.

S. No.	Method of exposure of nutrient agar plate	No. of colonies	
		Bacteria	Fungi
1			
2			
3			
4			
5			

S. No.	Source of inoculum	Growth
1		
2		
3		
4		
5		

10.2. Bacteriological Quality Testing of Water

For testing the bacteriological quality of potable water, bacteriologists test the harmless fecal bacteria, which are always present in the intestine. These bacteria include *Escherichia coli* type 1, *Streptococcus faecalis* and *Clostridium welchii* and are known as fecal indicator bacteria. Their presence in water indicates fecal contamination, which means that the water under test may prove to be a potential health hazard because it may contain enteric pathogens like *Salmonella* and *Shigella.*

The presence of coliforms is demonstrated by the use of a selective media in **Presumptive coliform test**. The selective media used is MacConkey's medium that contains bile salts inhibitory for the growth of non-intestinal lactose fermenting bacteria. The test is called presumptive because MacConkey's broth inoculated with water may give a positive test due to non-coliform organisms. Since a statistical method is employed to estimate the coliforms, the results are expressed as most probable number of coliforms or MPN. The MPN is calculated by consulting a statistical table (Table 10.1).

Objectives:

1. To examine the bacteriological quality of potable water by MPN count technique.
2. To ascertain the effectiveness of water purification process.

Materials required: Double strength MacConkey's broth tubes containing inverted Durham's tube (10 ml), Single strength MacConkey's broth tubes with inverted Durham's tube (10 ml), Sterile pipettes (1 & 10 ml), Water sample.

Procedure:

1. Shake the bottle containing water thoroughly to mix the contents.
2. Remove the stopper and aseptically transfer 10 ml of water sample into three tubes containing 10 ml of double strength MacConkey's broth using 10 ml pipette. Ensure that the pipette and neck of the sample bottle are flamed properly between inoculations.
3. Using 1 ml pipette inoculate 1 ml of water sample into three tubes containing 10 ml of single strength MacConkey's broth.

Table 10.1. Determination of most probable number by multiple tube test

Number of tubes giving positive reaction out of			MPN index per 100 ml
3 of 10 ml each	3 of 1 ml each	3 of 0.1 ml each	
0	0	1	3
0	1	0	3
1	0	0	4
1	0	1	7
1	1	0	7
1	1	1	11
1	2	0	11
2	0	0	9
2	0	1	14
2	1	0	15
2	1	1	20
2	2	0	21
2	2	1	28
3	0	0	23
3	0	1	39
3	0	2	64
3	1	0	43
3	1	1	75
3	1	2	120
3	0	2	93
3	2	1	150
3	2	2	210
3	3	0	240
3	3	1	460
3	3	2	1100

From: Standard methods for examination of water and waste water, 14th edition. American Public Health Association, Washington, D.C. 1975

4. Inoculate three tubes containing 10 ml of single strength MacConkey's broth with 0.1-ml water sample using 1-ml sterile pipette.
5. Incubate all the tubes at 37°C for 48 hours.

Results and observations: Change of colour from red to yellow indicating production of acid and gas bubble in the Durham's tube is a positive test.

Water sample No. and source	Number of tubes giving positive reaction out of			No. of coliforms present in 100 ml of water (MPN)
	3 Test tubes (10 ml)	3 Test tubes (1 ml)	3 Test tubes (0.1 ml)	
1				
2				
3				
4				
5				
6				
7				

10.3. Standard Plate Count of Water—A Quantitative Test

Testing of bacteriological quality of water is essential to prevent transmission of water borne diseases of fecal origin. A variety of microorganisms are present in water, which differ in their physiological requirements. Some of these microorganisms may be pathogenic. Though it may not be feasible to detect and isolate pathogens from water because of their small number, the total number of microorganisms present in a unit volume (1-ml) will certainly indicate the effectiveness of water purification process.

Objectives: To examine the quality of potable water by determining its microbial load.

Materials required: Tryptone glucose agar, Sterile Petri plates, Sterile pipettes (1 ml), Sterile dilution blanks (99 ml), Water sample.

Procedure:

1. Liquefy tryptone glucose agar medium by heating, cool to 50°C and keep in a water bath set at 50°C to maintain the medium in molten form.
2. Take two 99-ml dilution blank and prepare dilutions of 1:100 and 1:1000.
3. Label two plates each of 1:100, 1:1000, and 1:10000 dilutions.
4. Aseptically transfer aliquots from the dilutions prepared earlier to obtain the dilutions marked on Petri plates.
5. Pour 20-25 ml tryptone glucose agar in each of the plate, mix the diluted sample with the medium by gently rotating the plate. Allow to solidify the agar.
6. Incubate at 30°C for 24 hours.

Results and observations: Count number of colonies from plates having 30-300 colonies and record your observations.

Sample No. and source	Average No. of colonies per plate	Dilution	Number of organisms ml^{-1}
1			
2			
3			
4			
5			

10.4. Microbial Population Counts of Soil

Soil is a store house of numerous types of microorganisms including bacteria, fungi, protozoa and viruses. These microorganisms degrade organic waste materials (proteins, nucleic acids) like animal wastes and dead tissues of plants and animals into simple inorganic compounds (nitrite, nitrate, CO_2) and thus replenish the fertility of soil. Thus microbial flora of soil play a vital role in recycling of elements like carbon, nitrogen, phosphorus, sulfur etc. and help in the sustenance of life on this planet. Because of great diversity in the microbial flora of soil, no single technique can be used to count all types of microorganisms in a given soil sample. However, number of bacteria, fungi or any other microorganism can be counted by using specific growth medium.

Objectives:

1. To familiarize the students with the distribution and abundance of microorganisms in soil.
2. To determine the number of bacteria, actinomycetes and molds in a sample of soil.

Materials required: Nutrient agar, Glucose peptone acid agar, Glycerol yeast extract, Sterile dilution blanks (99 ml), Sterile Petri plates, Sterile pipettes (1 ml), Soil sample.

Procedure:

1. Liquefy the three types of media, cool to 50°C and maintain in a water bath set at 50°C.
2. Label Petri plates according to type of organism and dilutions. For different organisms you will use the following dilutions:

Bacteria	Actinomycetes	Molds
1:10000	1:1000	1:100
1:100000	1:10000	1:1000
1:1000000	1:100000	1:10000
1:10000000	1:1000000	1:100000

3. Label three 99-ml sterile dilution blanks A, B and C.
4. Weigh one gram of soil and add to dilution blank A. Shake vigorously for 5 minutes on a rotary shaker, if possible. Carry out serial dılutions.
5. Distribute appropriate amount of water from dilution blanks to respective Petri plate.
6. Pour appropriate medium to Petri plates, rotate the plates gently and allow them to cool.
7. Incubate the plates in incubator at 25°C for 3 to 7 days.
8. Count the number of colonies.

Results and observations:

Organisms	Count per plate (a)	Dilution (b)	Organisms per gram of soil (a × b)
Bacteria			
Actinomycetes			
Molds			

10.5. Isolation of an Antibiotic Producer from Soil

Antibiotics are chemicals produced by one type of living cell and are inhibitory for the growth and survival of other non-producing cells. There are many such agents that inhibit specific metabolic processes in microbial cells without affecting the metabolism of the host organism because of basic difference in the cellular organisation of Procaryotes (bacteria) and Eucaryotes (host). Such chemicals are useful in chemotherapy. Many soil organisms including bacteria, molds and actinomycetes are known to produce antibiotics, however, only a few of them are suitable for medical use. Research institutions and pharmaceutical companies are engaged in search for new microbes that can be used for production of new antibiotics.

Several compounds synthesized in laboratory are also capable of inhibiting growth of microorganisms. Such chemicals are known as **Synthetic antimicrobials**. A chemically modified antibiotic depicting wide spectrum and enhanced activity is termed as **Semi-synthetic antibiotic**.

Objectives: To isolate an antibiotic producing *Actinomyces*.

Materials required: Plates containing sterile glycerol yeast extract agar, Trypticase soy agar, Sterile Petri plates, Saline dilution blank (90 ml), Saline dilution blanks (9 ml), Sterile pipettes (1 ml), Alcohol, Spreader, *Staphylococcus epidermis* culture.

Procedure:

1. Weigh accurately 10 gram of soil and put into 90 ml saline dilution blank. Keep on a rotary shaker for 10-15 minutes to disperse the soil throughout the normal saline.

2. Label five 9 ml dilution blanks 1 through 5 and make tenfold dilutions by transferring 1 ml from diluted soil (90 ml blank) to 9 ml blank and further to other 9 ml blanks. Mix thoroughly before each transfer and use fresh pipettes for each transfer.
3. Label glycerol agar plates with dilution 1:10000, 1:100000 and 1:1000000 and transfer one ml diluted soil suspension from the respective tubes.
4. Spread the diluted soil suspension over the agar surface using L-shaped glass spreader. Sterilise the glass spreader every time by dipping in alcohol and flamming. The spreader should be cooled before use.
5. Incubate the plate for 7 days.
6. After 7 days, liquefy trypticase soy agar, cool to 50°C and mix into it 10 ml. *S. epidermis* broth culture. Pour the contents into sterile Petri plate and allow to solidify.
7. Examine the plates of glycerol yeast extract agar inoculated in the previous period for the presence of *Actinomyces* like colonies. *Actinomyces* colonies are generally embedded into agar and have a dusty appearance because of spore formation.
8. Scrap spores of *Actinomyces* like colonies with the help of sterile inoculation needle and streak on trypticase soy agar plates. Everytime pick and streak a different colony.
9. Incubate the plates at 30°C in an incubator for 7 days.
10. Examine the plates after 7 days and look for inhibition of *S. epidermis* growth. If antibiosis is present, transfer the spores of *Actinomyces* to fresh trypticase soy agar plates to prepare a pure culture.

Results and observations:

1. Describe the appearance of colonies.
2. Observe for presence of antibiosis and measure, if possible, the zone of inhibition of *S. epidermis* growth.
3. Compare the zone of inhibition of different colonies.

10.6. Determination of Anti-microbial Spectrum of Soil Isolates

Objective: To screen the soil isolates for their antimicrobial spectrum

Materials required: Trypticase soy agar, Cultures of *Escherichia coli*, *Staphylococcus aureus*, *Pseudomonas auriginosa*, *Mycobacterium smegmatis*, Sterile Petri plates.

Procedure:

1. Label the trypticase soy agar plates with the name/code number of soil isolate.
2. Mark a single line on the underside of the plate with glass marking pencil so as to divide the plate in two equal halves.
3. Make single line streak inoculation of antibiotic producing soil isolate on the marked line aseptically and incubate for 3-5 days in inverted position at 25-30°C.

4. After incubation, draw 4 lines perpendicular to the growth of the antibiotic producing soil isolate.
5. Aseptically make single line streak inoculation of each of the four test cultures. Start close to the growth & streak towards the edge of the plate. Do not touch the growth of antibiotic producing isolate.
6. Incubate the plates in inverted position for 24 hrs. at 37°C.

Observations and results: Examine all the plates for inhibition of growth of test organism and record the presence (+) or absence (–) of antibiotic activity against each of the test organism.

Soil Isolate	Test Organism				Spectrum
	E. coli G –ve	*S. aureus* G +ve	*P. auriginosa* G –ve	*M. smegmatis* Acid fast	
Unknown Isolate					
Positive Control					

10.7. Isolation of Nitrogen Fixing Bacteria

Although 80 per cent of the earth's atmosphere is molecular nitrogen (N_2), a majority of organisms (except a few bacteria) cannot use nitrogen in this form. Some bacteria and cyanobacteria termed **Nitrogen-fixing** microorganisms convert atmospheric nitrogen to ammonia, which can then be converted to organic nitrogen (proteins and nucleic acids), by plants. The nitrogen fixing microorganisms produce an enzyme known as **Nitrogenase** which enables the organism to combine atmospheric nitrogen with hydrogen to form ammonia at ambient temperature and pressure.

There are many different types of bacteria that fix atmospheric nitrogen, however, most important nitrogen fixers belong to two families, **Azotobacteriaceae** and **Rhizobiaceae**. The members of Aztobacteriaceae fix nitrogen independently and are known as **Asymbiotic** and those belonging to family Rhizobiaceae can fix nitrogen only in association with leguminous plants and are termed as **Symbiotic**.

A. Azotobacteriaceae

There are two genera in family Azotobacteriaceae that fix nitrogen as free living organisms under aerobic conditions. The basic difference in these genera is that *Azotobacter* produce drought resistant cysts but *Azomonas* does not. Both the genera possess Gram negative motile rods that may be oval or sometimes spherical in shape. Catalase is produced by both genera.

Table 10.2. Some distinguishing characteristics of Azotobacteriaceae

	Cysts	Motility	Water soluble pigment				
			Brown black	Brown black to Red violet	Red violet	Yellow green (Fluorescent)	Blue white (Fluorescent)
Azotobacter chroococcum	+	+	–	–	–	–	–
A. vinelandii	+	+	–	–	a	a	+
A. beijerinckii	+	–	–	–	–	–	–
A. nigricans	+	–	a	+	a	–	–
A. armaniacus	+	+	–	+	+	–	–
A. paspli	+	+	–	–	+	–	a
Azomonas agilis	–	+	–	–	–	–	+
A. insignis	–	+	b	–	a	–	a
A. macrocytogenes	–	+	–	–	–	–	a

a = 11-89% positive, b = 11-89% positive on benzoate

There are six species of *Azotobacter* and three of *Azomonas.* Some other characters of the two are given in Table 10.2.

Objectives: To isolate free living-nitrogen fixing bacteria by making use of selective media.

Materials required: A bottle containing 50-60 ml Burk's nitrogen free medium, Sterile Petri plates, Burk's nitrogen free agar, Glass slides, Cover slips, UV lamp, Garden soil sample.

Procedure:

1. Mix the soil sample thoroughly and transfer approximately one gram soil into bottle containing Burk's nitrogen free medium using a small spatula. Screws cap the bottle and shake the bottle to mix the soil and the medium.
2. Incubate the bottle at 30°C in horizontal position for 4-7 days. Loosen the cap of bottle slightly before incubation.
3. After incubation for 7 days prepare a wet mount without disturbing the soil at the bottom.
4. Examine the slide using a high and dry objective. If available use a phase contrast microscope. Look for large ovoid to rod shaped organisms either as singles or in pairs.
5. Record the motility of *Azotobacter* like organisms.
6. Try to look for cysts also. Cysts resemble endospore and are refractile. However, cysts may not be present since it takes two week to form cysts.
7. If *Azotobacter* like organisms are present, streak a loopful on a nitrogen free agar plate.

8. Incubate the plate at 30°C for 4-5 days.
9. After incubation of 4-5 days examine the plate in ultraviolet light and record the observations.

Observations:

1. Note if the colonies were pigmented.
2. If pigmented note the colour.
3. Look for fluorescence in UV light, if positive note the colour.
4. Prepare a slide and note Gram reaction.

B. Rhizobiaceae

Symbiotic nitrogen fixers of genera *Rhizobium* and *Bradyrhizobium* are the principal nitrogen fixers and their contribution towards enrichment of soil is more compared to free-living nitrogen fixers. Members of these genera when present in soil induce nodulation on the roots of specific leguminous plants and fix atmospheric nitrogen in association with them. This is a type of mutualism where both the partners are benefitted. Non-leguminous plants of many types can set up similar symbiotic association with members of genus *Frankia.* Since nodules contain specific rhizobia, nodules from specific host plant are used for isolation of *Rhizobium* or *Bradyrhizobium.*

Objectives:

1. To collect nodules of leguminous plants.
2. To isolate *Rhizobium* from the nodules.

Materials required: Yeast mannitol agar, Sterilised Petri plates, Calcium hypochlorite (3%), Ethanol (95%), Sterilised distilled water, Forceps, Spade.

Procedure:

1. Select one representative plant of each legume for sampling.
2. Mark a circle with a radius of approximately 15-20 cm with a spade and cut out this section approximately 20 cm deep.
3. Using the spade slowly lift the plant alongwith clump of soil.
4. Remove excess soil from the roots with your hands without damaging secondary roots as the nodules present on the lateral roots may get lost.
5. Gently place the whole plant in a plastic bag and label.
6. Wash the roots thoroughly to remove soil in a running tap water. Alternately dip the plant along with soil clump in a bucket full of water and remove the soil sticking to the roots gently.
7. Now put the root sample in a sieve and wash the roots once again under a gentle stream of water from a tap.
8. Remove the nodules from the root by cutting the root about 0.5 cm on each side of

attachment of nodule. About 10 healthy and pink nodules may be collected from each plant.

9. Put the intact nodules in a sterilised Petri plate and add 95% ethanol. Allow it to stand for about 30 seconds. This help in breaking the surface tension and remove air bubbles from the tissue.
10. Using a sterile forceps, transfer the nodules to a 3% solution of calcium hypochlorite in a sterile Petri plate and allow to stand for 3-5 minutes. Calcium hypochlorite sterilises the outer surface of nodules.
11. Remove the nodules with a forceps and rinse in five changes of sterile water to remove the traces of calcium hypochlorite from the nodule surface. A 0.1% solution of mercuric chloride can also be used for surface sterilisation of nodules. However, mercuric chloride is highly toxic and should be avoided as far as possible.
12. Crush the surface sterilised nodule with a sterile glass rod in a drop of sterile water in a Petri plate.
13. Streak a loopful of pink sap on yeast-mannitol agar plates and incubate at 25-30°C in dark.
14. Repeat steps 12 and 13 each time taking another nodule.
15. After 4-10 days examine the plates and look for well isolated colonies.
16. Pick up a single colony typical of *Rhizobium* and perform a Gram-stain.

Results:

1. Well-isolated colonies will appear in 3-5 days in case of fast growing rhizobia (e.g. *Leucaena, Cicer, Pisum, Sesbania* etc.) and in 5-7 days in slow growers (e.g. *Glycine max*).
2. Colonies of rhizobia are usually discrete, round varying from flat to convex. The colonies on yeast mannitol agar are white-opaque or milky or watery-translucent.
3. Rhizobia do not absorb colour of congo red in dark. Some species may absorb colour of the dye when incubated in light.

Observations:

1. Look for colonies typical of *Rhizobium* and transfer to a yeast-mannitol slant.
2. Perform Gram-reaction, streak on yeast mannitol plates and obtain a pure culture.
3. Try to authenticate by performing plant infection test.

REVIEW QUESTIONS

1. How microorganisms are useful as decomposers?
2. How does microorganism exist everywhere?
3. How a microorganism isolated from Antarctic region will differ from that isolated from hot deserts?

4. What are indicator microorganisms? What do you understand by presumptive test?
5. What is enrichment culture?
6. What are antibiotics? How antibiotics differ from synthetic antimicrobials?
7. What is plant infection test?
8. What is Cyst? How does cyst differ from an endospore?
9. How *Azotobacter* differs from *Azomonas*?
10. Why some microorganisms produce antimicrobial substances?
11. Why Durham's tube is used in microbiological quality testing of water?
12. How will you demonstrate the ubiquitous presence of microorganisms?
13. How microbes are useful in sustenance of life on earth planet?
14. Why the colour of MacConkey's broth changes from red to yellow during growth of microorganisms?
15. What role the microorganisms play in the recycling of nutrients?
16. What is the need to degrade the organic form of nutrients into inorganic form?

Chapter 11

Bacterial Genetics

Microorganisms share many biochemical properties with cells of other multicellular organisms. This coupled with the ability of rapid growth and easy genetic manipulations make them excellent models for the understanding of principles behind living processes. The rapid rate of growth permits observation of transmission of a trait through many generations and the availability of large test populations that allows isolation of spontaneous and induced mutants. In addition the low cost of maintenance and propagation make it possible to perform a large number of experimental procedures. Thus microorganisms have proved to be essential tools in basic research in Biochemistry, Genetics, Biotechnology, Genetic engineering and Molecular biology.

All organisms including bacteria have a precise sequence of nucleotides in their DNA. This sequence of nucleotides is termed as **Genotype**. Occasionally there is a change in the sequence of nucleotides in the genome of a strain. Such a strain carrying a change in the nucleotide sequence is called a **Mutant**. Thus a mutant differs from its parent in genotype. Some mutants also depict changes in observable property i.e. **Phenotype** in addition to alteration in genotype. The parent strain isolated from nature is called **Wild** type and mutants are derived from wild type or another mutant. Mutations are mostly harmful although sometimes-beneficial changes do occur.

Mutations may be spontaneous or induced. **Spontaneous mutations** occur as a result of action of chemical and physical components in the organism's natural environment that alters the structure of bases in the DNA. Spontaneous mutations also occur during replication of DNA because of error in pairing. However such errors occur at a frequency of 10^{-7} - 10^{-11} per base pair per round of replication. The rate at which spontaneous mutations occur is also extremely low (10^{-7} per generation) in all organisms. **Induced mutations** are genetic changes resulting from the organism's exposure to physical or chemical agents commonly known as **Mutagens**. The resultant mutations are of the same types that occur spontaneously, however, the rate of mutation increases and in some cases dramatically so.

Mutations involving one base pair are known as **Point mutations** that arise as a result of the addition, deletion, or substitution of one or more bases in the region of a single gene. As a result, substitution of one or more amino acid occur during translation, and a protein that is synthesized may be inactive, reduced in activity, or entirely different. Some other types of mutations are Silent mutation, Miss-sense mutations and Frame shift mutations. Mutations are a mean of variation and diversity of microorganisms in nature.

Another source of genetic variability in nature is transfer of genetic material and its subsequent incorporation into bacterial genome through conjugation, transformation, transduction or plasmids. **Conjugation** is referred to a mating process between two 'sexually' differentiated bacterial strains that allows unidirectional transfer of genetic material ($F^+ \rightarrow F^-$. **Transformation** is a genetic alteration in a cell resulting from the introduction of free DNA from the environment across the cell membrane. **Transduction** is a process of genetic transfer in which bacterial DNA is transferred from one cell (the donor) to another (the recipient) encased within the capsid of a virus. Extrachromosomal, DNA molecules capable of independent existence and replication within the cell are termed as **Plasmid (s)**. The information encoded by plasmid DNA is active and contributes towards phenotype of their host cell. Plasmids have been found to be useful in the transfer of small fragments of DNA from one to another cell.

11.1. Isolation of Auxotrophic Mutants

A bacterial strain capable of growing on minimal medium in the absence of any growth factor is termed as **Prototroph** or "wild strain". On the contrary a mutant that requires one or more than one growth factor for its growth is called **Auxotroph**. Such mutants are deficient in the production of an enzyme required for the synthesis of essential organic compound or growth factor. Auxotrophic mutants that can not synthesize certain organic compounds such as amino acids, purine or pyrimidine bases or vitamins can be isolated.

There are two basic techniques for the isolation of mutants, the **Gradient plate** or **Direct selection** and **Indirect selection.** In this exercise indirect selection using **replica plating** shall be used for the isolation of auxotrophic mutants.

In replica plating, a culture is treated with a mutagen and bacteria are allowed to grow on complete medium like nutrient agar after which an imprint of the colonies is made on velveteen covered blocks and colonies are transferred on complete solid medium and on glucose-minimal salts solid medium. The colonies that appear on complete medium but not on the minimal medium are of auxotrophs for one or another organic compound. Auxotrophs are identified indirectly because they will not grow on minimal medium.

Objectives:

1. To define the terms mutation, prototroph and auxotroph.
2. To isolate auxotrophic mutants of bacteria by replica plating.

Material required: Exponentially growing culture of *Escherichia coli* or *Rhizobium*, Petri plates containing complete medium (Nutrient agar), Petri plates containing glucose-minimal salt medium, Ultraviolet lamp (260 nm), Sterile water blanks (1 & 99 ml), Sterile pipettes (1 ml), Spreading rod, Alcohol, Sterile velveteen cloth and Sterile replica plating block.

Procedure:

1. Transfer aseptically 1 ml of the broth culture into 99-ml water blank and mix well.
2. Take another pipette and transfer 1 ml from water blank 1 to another 99-ml water blank and mix.
3. Transfer 1 ml from water blank 2 to 99-ml water blank. 3 and 1 ml onto the surface of Petri plates containing nutrient agar.
4. Mix the contents of water blank 3 and transfer 1-ml and 0.1 ml to the surface of Petri plates containing nutrient agar separately.
5. Take the spreading rod, disinfect by dipping in alcohol and ignite the alcohol on the flame, cool the spreading rod after the alcohol burns off and spread the liquid on the surface of plates of nutrient agar uniformly starting from the highest dilution. Allow the plates to sit undisturbed until the liquid diffuses into the agar.
6. Remove the cover of the plates with agar side up and expose to ultraviolet light for 30, 40, 50 & 60 seconds at a distance of about 30 cm.
7. After incubation of the plates at 25°C for 24-48 hr. select plates with 25 to 50 isolated colonies and mark the plates with a reference mark. This is the master plate.
8. Also mark the uninoculated nutrient agar and minimal media plates with reference mark.
9. Pick up a piece of velveteen clothes from the Four Corners and secure them tightly on the block with rubber band.
10. Keep the replica-plating block on the table with velveteen surface up. Invert the master plate selected in step vii on the block so as to touch the velveteen lightly.
11. Remove the cover of the minimal medium, align the reference mark and touch the uninoculated minimal agar plate with the inoculated replica-plating block. Replace the cover. Repeat the same step using uninoculated complete medium plate.
12. Refrigerate the master plate and incubate the replica-plated plates at 25°C for 24-48 hr.
13. After incubation, compare the plates for growth on complete and minimal media.
14. Pick the colonies that don't grow on minimal medium but grow on complete medium.
15. Pick such colonies from the master plate, purify and check for auxotrophic properties.

Results and observations: Enumerate the number of colonies of auxotrophs on plates exposed to UV for different period and record.

Bacterial Culture:		
UV Exposure (Sec.)	Number of Auxotrophs Dilution	
	10^{-3}	10^{-4}
30		
40		
50		
60		

11.2. Isolation of Streptomycin Resistant Mutants

Many a times a microbial cell survive in otherwise deleterious environment. An example of such a genetic adaptation is the development of antibiotic resistance in a small population of microorganisms. The development of antibiotic resistance in microorganisms is of major clinical importance because the number of drug resistant microbial strains continues to increase due to their extensive use and frequent misuse, which over the years have resulted in the development of drug resistant strains. These agents select for the resistant mutant and do not act as inducers.

In a drug resistant organism, the mutated gene enables the cell to circumvent the antimicrobial effect of the drug by a variety of mechanisms, such as:

- The production of an enzyme that alters the chemical structure of the antibiotic, thus rendering it ineffective. e.g. penicillin resistance.
- A change in the permeability of the cell membrane e.g. streptomycin resistance.
- A decrease in the sensitivity of their enzymes to inhibiting mechanisms, as in the resistance to streptomycin, which interferes with the translation process at the level of ribosomes.
- An overproduction of a natural substrate (metabolite) to compete effectively with the drug (antimetabolite), as in the case of resistance to sulfonamides, which produce their antimicrobial effect by competitive inhibition.

Streptomycin-resistant mutant from a prototrophic (wild type, streptomycin sensitive) *Escherichia coli* culture can easily be isolated by means of the gradient plate technique. In this technique a double layered agar plate is prepared. The lower, slanted agar medium layer lacks streptomycin while the upper layer of the molten agar medium containing the antibiotic produces a streptomycin concentration gradient in the surface layer. When prototrophic test culture is spread over the surface of agar medium and incubated, the appearance of colonies in a region of high streptomycin concentration indicates the development of streptomycin resistant mutants.

Objectives: To isolate streptomycin-resistant mutants in a prototrophic bacterial population by means of the gradient-plate technique.

Materials required: Fresh nutrient broth cultures of *Escherichia coli,* Trypticase soy agar deep tubes (10-ml), Streptomycin solution (10 mg per 100 ml of water), Sterile Petri dish, Sterile pipettes (1ml), Bent glass rod, Beaker, 70% Ethanol, Water bath, and Glassware marking pencil.

Procedure:

1. Melt two trypticase soy agar tubes, cool and maintain at 45°C in a hot water bath.
2. Place a sterile Petri dish in tilted position by keeping a glass rod/pencil beneath one end. Pour into it sufficient amount of the molten trypticase soy agar medium to cover the entire bottom surface, and allow to solidify in the slanted position.
3. Using a sterile 1-ml pipette, add 0.1 ml of the streptomycin solution to a second tube of molten trypticase soy agar. Mix by rotating the tube between the palms of your hands.
4. Place the dish prepared in step ii in a horizontal position and pour in a sufficient amount of the molten agar medium containing streptomycin to cover the gradient agar layer and allow to solidify.
5. Add 0.2 ml of the *E. coli* test culture aseptically and spread the culture over the entire agar surface with the help of an alcohol dipped and flamed bent glass rod.
6. Incubate the plates for 48 hours at 37°C and select one or two isolated colonies present in the middle of the streptomycin concentration gradient.
7. Using a sterile inoculating loop pick the selected colonies and streak toward the high-concentration end of the plate.
8. Incubate the plates in an inverted position for 48 hours at 37°C.

Results and observations:

1. Observe the plates for the appearance of streptomycin resistant colonies and note their position on the surface of plates.
2. After the second incubation, observe for the presence of colonies into the area of high streptomycin concentration. Growth in this area indicates the presence of streptomycin resistant mutants.

11.3. Enzyme Induction

Each bacterial cell is capable of synthesising hundreds of different enzymes. Studies have shown that these enzymes are not present within the cells in equal concentrations always. Some enzymes, called **Constitutive enzymes**, are synthesised at a constant rate regardless of conditions in the cell's environment. However some enzymes are synthesised by the cell only when necessary, and are subject to regulatory mechanisms. One such mechanism, induction, requires the presence of specific substrate that acts as the inducer. Such enzymes are known as **Inducible enzymes**. An extensively studied inducible enzyme in *Escherichia coli* is **β-galactosidase**, which acts on the disaccharide lactose to yield the monosaccharides glucose and galactose. The gene for β-galactosidase is a member of the lactose operon that functions as a unit, all being transcribed only when the inducer, lactose, is present in the surrounding.

To illustrate β-galactosidase induction, two test strains of *E. coli* are used, a prototrophic (wild type) strain (*lac* Z^+) and an auxotrophic (mutant) strain (*lac* Z^-), which carries a mutation in the gene for β-galactosidase as well as a mutation in the lactose operon regulatory gene.

Orthonitrophenyl-β-D-galactoside (ONPG), a colorless analog of lactose, can serve as the substrate for the induction of β-galactosidase synthesis. After induction, β-galactosidase hydrolyses ONPG to galactose and a yellow nitrophenolate ion and is indicated by the appearance of a macroscopically visible yellow color in the medium. Absence of this color change indicates that enzyme induction did not occur as is observed in the lactose negative strain.

Objectives: To illustrate experimentally the induction of β-galactosidase synthesis in *Escherichia coli.*

Materials required: Inorganic synthetic broth suspensions of 12 hour nutrient agar cultures of a lactose-positive (*lac* Z^+) *Escherichia coli* strain and a lactose-negative (*lac* Z^-) *Escherichia coli* strain (25 ml each), Sterile 10% glucose, 10% lactose, Toluene and o-nitrophenyl-β-D-galactoside (ONPG). Sterile pipettes (1 & 5 ml), Sterile 13 × 100-mm test tubes, Sterile 25-ml Erlenmeyer flasks, Spectrophotometer, Shaking water-bath incubator, and Glassware marking pencil.

Procedure:

1. Grow both test strains in the following media for 12 hrs.:
   (a) Inorganic synthetic medium lacking an organic carbon and energy source that is required by the heterotrophic *E. coli.*
   (b) Inorganic synthetic medium containing glucose, which can be utilized by both strains as a carbon and energy source.
   (c) Inorganic synthetic medium containing lactose, which can be utilized only by the prototrophic strain.
2. Adjust the absorbance to 0.1 at 600 nm.
3. Label three sterile test tubes and three sterile 25-ml Erlenmyer flasks as '*lac*$^+$' and the name of the substrate to be added (glucose, lactose, or water). Similarly label three sterile flasks '*lac*$^-$' for each organism.
4. Using sterile 5-ml pipettes, aseptically transfer 5 ml of the *lac*$^+$ and *lac*$^-$ inorganic synthetic broth cultures to their respectively labeled test tubes.
5. Aseptically add 0.5 ml of the glucose and lactose solutions and 0.5 ml of sterile distilled water to the appropriately labeled tubes.
6. Determine the absorbance of all cultures at a wavelength of 600 nm.

7. Aseptically transfer each culture to its appropriately labeled flask.
8. Incubate all flasks for 2 hours in a shaking water-bath at 37°C.
9. Following incubation, transfer all cultures back to their appropriately labeled test tubes and determine the absorbance for each culture at a wavelength of 600 nm.
10. To each culture, add 5 drops of toluene and shake vigorously (toluene ruptures the cells, releasing intracellular enzymes and permitting the entrance of ONPG).
11. To each culture, add 5 drops of ONPG solution and incubate for 40 minutes at 37°C in a shaking water bath.
12. Observe all cultures for development of yellow color and note the time. Determine the absorbance of yellow colour at 400 nm.

11.4. Isolation and Detection of plasmid (s) in *Escherichia coli*

Plasmids are self-replicating small DNA molecules with covalently closed circular (ccc) configuration. Plasmid are not essential for survival of bacteria but may encode a wide variety of genetic determinants that permit the bacteria to survive better in adverse environmental conditions. Plasmid DNA has same structure and properties as that of chromosomal DNA. Plasmids are found in a variety of bacteria with diverse functional genes. For example Ti plasmid in *Agrobacterium tumifaciens* codes for tumor formation. In *Rhizobium*, the genes for nodulation are present on plasmids. Plasmids vary in size, the smallest being 1 kb to largest containing 1700 kb. The techniques of isolation of plasmids require means for gentle lysing of bacterial cells to preserve the plasmid DNA intact to distinguish it from chromosomal DNA. For detection of plasmids three method are commonly used:

Physical: The technique used is agarose gel electrophoresis in which the cells are lysed and plasmid DNA is detected by Polyacrylamide Gel Electrophoresis.

Transfer: If a bacterial trait is expected to be plasmid borne, gene transfer will document the presence of plasmid (s). The method involves the transfer of plasmid (s) carrying genes for antibiotic resistance from antibiotic resistant strains of bacteria to antibiotic sensitive strains. The presence of resistance to antibiotic in recipient strain indicates the transfer and presence of plasmid.

Curing: The plasmids can be removed by treatment with chemicals like ethidium bromide, or acridine orange. Some microorganisms are known to lose plasmids when grown at elevated temperatures. The elimination of genetic trait, determined by plasmid (s) by exposure to curing agents suggests the presence of plasmid (s). Plasmids carrying genes for resistance to antibiotics have successfully been removed in *E. coli* and other bacteria sensitive to the drug.

Some bacteria possess chromosomal as well as plasmid DNA secured by cell wall and cytoplasmic membrane. Cell wall can be degraded by treatment with enzyme lysozyme and the cytoplasmic membrane is broken by the use of a detergent. Once DNA is in solution, it has negative charge due to presence of phosphate groups. If a current is applied in solution having negatively charged DNA, it moves towards anode. For separating the plasmid DNA gelling agent such as agarose is used. It is a purified form of agar and is a polymer of galactose and galactose only. The gel is loaded with sample and electric current is applied to separate plasmid DNA on the basis of size. The migration of different DNA species is related inversely to their molecular weights i. e. the larger the molecular weight the slower is the rate of migration.

When the size of plasmid is large, isolation of plasmid by normal methods is not feasible because the plasmid degrades. For such plasmid *in situ* gel lysis is used. The cells are lysed inside the well under very gentle conditions. This method is useful for detection of plasmids in *Rhizobium* and *Agrobacterium* that possess large plasmids.

A. Isolation of Plasmid DNA

Objectives: To demonstrate the presence of plasmid (s) in *Escherichia coli* containing plasmids of different sizes.

Materials required: Fresh culture of *Escherichia coli*, TRIS-EDTA buffer (pH 8.0), TES buffer, Lysozyme.

Triton X-100 lytic mixture:

10% Triton X-100 in 0.01 M TRIS (pH 8.0)	1 ml
0.25 M EDTA (pH 8.0)	25 ml
1 M TRIS (pH 8.0)	5 ml
Water	69 ml

Procedure:

1. Inoculate *Escherichia coli* in a suitable medium preferably in the presence of antibiotics to prevent the loss of plasmids during growth.
2. Incubate the flask on rotary shaker at 37°C with gentle aeration (100-120 rpm).
3. Harvest the cells by centrifugation of 10-20 ml broth culture.
4. Discard the supernatant and suspend the pallet into 1-2 ml of TES buffer and centrifuge for 10 min. at 10000 rpm at 5°C.
5. Resuspend the pallet in 200 l ice-cold 25% sucrose solution in 0.05 M TRIS and 0.001 M EDTA (pH 8.0).
6. Place the tube in ice for 30 min. and add 20 l lysozyme (5 mg ml^{-1} in 0.25 M TRIS, pH 8.0). Mix the contents of the tube and place in ice for 10 min.
7. Add 40 μl of 0.25 M EDTA (pH 8.0), swirl the tube and place in ice for another 10 min.
8. Add 100-200 μl of Triton-X lytic mixture and again place in ice for 20 min. after gentle swirling.
9. Centrifuge at 35000 g for 20 min at 5°C.
10. About 95% of plasmid DNA is separated from chromosomal DNA which, can be

further purified and concentrated by Cesium chloride–ethidium bromide density gradient centrifugation.

B. Detection of Plasmid (s) by *in situ* Gel Electrophoresis

Objectives: To demonstrate the presence of plasmid (s) in *Escherichia coli* containing plasmids of different sizes.

Materials required: Fresh culture of *Escherichia coli,* Agarose, Tris-acetate buffer (TAE), TRIS-borate buffer (TBE), TRIS-EDTA buffer (TE, pH 8.0), TES buffer, Lysozyme, RNase (2 mg ml^{-1} in acetate buffer of pH 8.8), Ethidium bromide (10 mg ml^{-1}).

Lysis mixture: Prepare TE buffer (TRIS-50 mM, EDTA-10 mM, Sucrose - 8%,. Ficoll - 2%) and autoclave. Add lysozyme @ 1 mg ml^{-1} prior to use.

Procedure:

1. Grow *E. coli* culture having plasmids of different size and carrying antibiotic marker. The bacterium should be grown in the presence of antibiotics to prevent the loss of plasmid (s) during growth.
2. From overnight grown culture inoculate again in LB broth to get a culture suspension of approximately 0.2 - 0.3 O. D. at 600 nm. Such cells will be in log phase.
3. Prepare agarose gel simultaneously using 0.8% agarose. However for the detection of large sized plasmid use lower concentration of agarose. The gel is prepared in buffer containing detergent (TAE and TBE + 0.6 ml of 10% SDS, for breaking cells). Allow it to solidify.
4. High concentration of chromosomal DNA interferes in the detection of plasmid DNA. Thus it is advisable to use low concentration of cells.
5. Take 200 µl of cell suspension and centrifuge at 8000 rpm for 3 minutes. Discard the supernatant and resuspend the pellet in 10 µl of TRIS-EDTA Buffer (pH 8.0).
6. Keep the tube on ice and add 50 1 of lysis mixture containing sucrose 8%, ficoll, lysozyme and RNase.
7. Load the mixture immediately into wells (25 µl). Run gel at 10-15 v for 10-15 minutes.
8. Apply 80-100 V current for 2 hrs. During this period plasmid DNA come out due to small size while chromosomal DNA remain in the well.
9. The gel is stained after removing SDS to visualize plasmid. To remove SDS, place the gel in running water or change with fresh water 2-3 times prior to staining. Now stain the gel with ethidium bromide that intercalates between 2 bases and when examined under UV light gives fluorescence.

11.5. The Ames Test: A Bacterial Test System for Chemical Carcinogenicity

Our exposure to a wide variety of chemical compounds has increased markedly over the past decades, some of which may be carcinogenic. It is believed that the intrusion of these chemicals in the form of industrial pollutants, pesticides, food additives, hair dyes, cigarette smoke etc. may play a significant role in the induction of malignant transformations in humans. There is strong evidence that links carcinogenicity to mutagenicity. Research indicates that approximately 90% of the chemicals cause cancer by inducing mutations in somatic cells. These mutations are most frequently due to substitution of one base for another in the DNA molecule resulting in frameshift mutations, a shift in the reading frame of a gene because of the addition or deletion of a base.

In view of the rapid advent of new products being thrown into market, it is essential to determine their potential genetic hazards. In the past, animal models have been used to evaluate the chemicals for their carcinogenic potential. However, the procedures are not only costly but time consuming also. Since the chemical nature of DNA of bacteria is similar to that of other organisms, the use of bacterial test systems for the rapid detection of possible mutagens and their carcinogenic effect may prove economical. The **Ames test** is a simple and inexpensive procedure developed by **Bruce Ames** in which a bacterial test organism is used to screen new chemicals for their mutagenicity. The test organism employed is a histidine-negative (his^-) and biotin-negative (bio^-) auxotroph of *Salmonella typhimurium* that will not grow on a medium deficient in histidine unless a back mutation to histidine-positive (his^+) has occurred.

The enzymatic pathways of an organism, whereby nonmutagens are transformed into mutagens and vice versa when introduced into human systems, frequently influence mutagenic effect of a chemical. This toxification or detoxification frequently occurs in the liver in mammals. Thus in Ames test generally a liver homogenate, S-9, which serves as a source that activates enzymes, to make this bacterial system more comparable to a mammalian test system is added. It has been established that many potent carcinogens are not directly carcinogens or mutagenic. However, such chemicals undergo changes, primarily in liver, where enzymes (mixed function oxygenases) normally involved in detoxification cause formation of expoxides or other activated forms of compounds, which are highly reactive with DNA

In the Ames test, molten agar containing the test organism, S-9 mix, and traces of histidine to allow the bacteria to undergo several cell divisions necessary for mutation to occur is poured on a minimal agar plate. A disc impregnated with the test chemical is then placed in the center of the test plate. Following diffusion of the test compound from the disc, a concentration gradient of the chemical is established. A qualitative indication of the mutagenicity of the test chemical can be determined by noting the number of colonies present on the plate. Each colony represents a $his^- \rightarrow his^+$ revertant. An increase in the number of colonies in the presence of test chemical as compared to the number of spontaneous revertants on the control plate indicates mutagenic effect of test chemical, a positive result.

A modified Ames test is described below where S-9 mix is not used to test for the mutagenicity of nitro compounds, which, as in humans, are activated by the bacterial nitro reductases. Four minimal agar plates are inoculated with the *S. typhimurium* test organism. One plate, the negative control, is not exposed to a test chemical. Colonies developing on this plate are representative of spontaneous $his^- \rightarrow his^+$ mutations. The second plate, the positive control, is exposed to a known nitrocarcinogen, 2-nitrofluorene. The remaining two plates are used to determine the mutagnenicity of two commercial hair dyes.

Objectives: To screen for potential chemical carcinogens using a bacterial test system.

Materials required: Nutrient broth cultures of *Salmonella typhimurium* his^- Strain, Minimal agar plates, 2-ml top agar tubes, Sterile biotin-histidine solution, 2-nitrofluorene dissolved in alcohol, Commercial hair dyes, Sterile pipettes (1 ml), Sterile discs, Forceps, Water bath, and Marking pencil.

Procedure: Wear disposable gloves and a laboratory coat when handling 2- nitrofluorene. Place excess of this chemical into a sealable container and put it inside a fume hood for subsequent disposal of hazardous materials.

1. Label minimum agar plates with the name of the test chemical to be used. Label another plate as a negative control.
2. Melt tubes of top agar in a hot-water bath and maintain the molten agar at 45°C.
3. To each molten top agar tube, aseptically add 0.2 ml of the sterile biotin-histidine solution and 0.1 ml of the *S. typhimurium* test culture and mix by rotating the test tube between the palms of your hands.
4. Aseptically pour the top agar cultures onto the minimal agar plates and allow to solidify.
5. Using sterile forceps, dip each disc into its respective test chemical solution and drain excess chemical by touching to the side of the container.
6. Place the chemical-impregnated discs in the center of the respectively labeled minimal agar plates. Place a sterile disc on the plate labeled negative control.
7. Gently press down on the discs with the sterile forceps, so that they adhere to the surface of the agar.
8. Incubate all plates in inverted position for 24 hours at 37°C.

Observations and results:

1. Count the number of large colonies on each plate and record.
2. Deduct the number of colonies on control plate from the number of colonies present on plates containing chemical mutagens, This represents the number of induced mutations.
3. Determine the relative mutagenicity of the test compound on the basis of induced mutations. Below 10, (–); 10, (1+); 100, (2+) and 500, (3+).

Test Chemical	Number of Colonies	Number of Induced Mutations	Degree of Mutagenicity
Control		–	–
2-Nitrofluorine			
Hair Dye 1			
Hair Dye 2			

REVIEW QUESTIONS

1. What is a mutant? How does it differ from a wild strain?
2. Differentiate between phenotype and genotype.
3. What do you mean by a mutagen? Name some mutagenic chemicals generally used in the laboratory for induction of mutations.
4. What do you understand by Frame shift mutations?
5. What is a point mutation?
6. Differentiate between constitutive and inducible -galactosidaseenzymes.
7. What is the function of ONPG in assay of enzyme?
8. Explain how genetic variations occur in nature?
9. What are the mechanisms of antibiotic resistance in microorganisms?
10. Why antibiotic gradient plate method is used for isolation of mutants?
11. There has been an increase in the drug resistance in bacteria in recent years, why?
12. Does antibiotic when added into medium cause mutations?
13. Explain the function of S-9 in Ames test.
14. What is the purpose of biotin-histidine solution in Ames test?
15. Explain the relationship in mutagenicity and carcinogenicity.
16. All carcinogens are mutagens but all mutagens are not carcinogens, Explain.
17. Why bacterial test systems are preferred over the mammalian systems to test for chemical carcinogenicity.
18. Explain the molecular mechanism of mutation.
19. Explain the mechanism of UV rays in the causation of mutations.
20. What is photoreactivation? How it differs from dark repair? How photoreactivation can be prevented?

Appendices

APPENDIX 1

Preparation of Stains and Reagents

Acid Fast Stain

Ziehl-Neelsen's Carbolfuchsin (hot stain)

10% Basic fuchsin solution in ethyl alcohol	10 ml
5% Aqueous phenol solution	100 ml

Mix, allow to stand overnight and filter

Kinyoan's Carbolfuchsin (cold stain)

95% Ethyl alcohol	25 ml
Basic fuchsin	3.5 g
Phenol	12.5 g

Dissolve by heating to 80°C. Add distilled water to make the volume to 300 ml. Add a few drops of Tween 80.

Acid Alcohol

Ethyl alcohol (95%)	97 ml
Hydrochloric acid (concentrated)	3 ml

Methylene Blue

Methylene blue	0.3 g
Distilled water	100 ml

Barritt's Reagent

Solution A

Alpha-naphthal	6 g
Alcohol (95%)	100 ml

Solution B

Potassium hydroxide	16 g
Distilled water	100 ml

Capsule Stain

Congo Red

Congo red	2 g
Distilled water	100 ml

Acid Fuchsin (Maneval's stain)

Phenol (5% aqueous solution)	30 ml
Glacial acetic acid (20% aquous solution)	10 ml
Ferric chloride (30% aquous solution)	4 ml
Acid fuchsin (1% aquous solution)	2 ml

Cleaning solution

Sodium or potassium Dichromate	120 g
Tap water	1000 ml
Sulphuric acid (concentrated)	1600 ml

Flagella Stain

Loeffler's Flagella mordant

20% Aqueous solution of tannic acid	100 ml
Ferrous sulphate	20 g
10% Basic fuchsin in ethyl alcohol	10 ml
Distilled water	40 ml

Dissolve ferrous sulphate in warm water and add other ingredients

Loeffler's Flagella stain

10% Basic fuchsin in ethyl alcohol	20 ml
3% Aniline water	80 ml

Gram Stain

Crystal violet

Crystal violet	2 g
Ethyl alcohol (95%)	20 ml
10% Aqueous solution of ammonium oxalate	80 ml

Gram's iodine

Iodine	1 g
Potassium iodide	2 g
Distilled water	300 ml

Ethyl alcohol (95%)

Ethyl alcohol (100%)	95 ml
Distilled water	5 ml

or

Acetone-Ethanol

95% Ethanol	70 ml
Acetone	30 ml

Safranine

2.5% Safranine in ethyl alcohol	10 ml
Distilled water	100 ml

Kovac's Reagent

p-Dimethylaminobenzaldehyde	3 g
Amyl or butyl alcohol	75 ml
Concentrated Hydrochloric acid	25 ml

Dissolve the aldehyde in the alcohol at 50-55°C. Cool and add acid. Store in a refrigerator in dark

Lacto-phenol Cotton Blue

Lactic acid	20 ml
Phenol	20 g
Glycerol	40 ml
Aniline blue	0.05g
Distilled water	20 ml

Dissolve the phenol crystals and lactic acid in hot water and then add aniline blue dye.

Loeffler's Methylene Blue

Methylene blue	0.5 g
1 % Potassium hydroxide	1 ml
Alcohol	30 ml
Water	100 ml

Warm water to 50 °C. Stir in methylene blue and add other ingredients. Filter

Methylene Blue Stain

Saturated solution of methylene blue in ethyl alcohol	10 ml
Distilled water	30 ml

Negative Stain

Nigrosin (water soluble)	10 g
Distilled water	100 ml

Keep in boiling water bath for 30 minutes. Add 0.5 ml formaline and filter

Newman-Breed Stain

95% Ethyl alcohol	40 ml
Tetrachloroethane	40 ml
Methylene blue	1 g
Glacial acetic acid	6 ml

Warm ethyl alcohol and tetrachloroethane to 70°C. Add methylene blue and dissolve by shaking. Cool and add acetic acid. Shake the mixture and store in dark.

Nitrite Test Reagents

Solution A	
N,N-Dimethyl-1 naphthyl-amine dihydrochloride	0.6 g
Acetic acid (5N)	100 ml
Solution B	
Sulfanilic acid	0.8 g
Acetic acid (5N)	100 ml

Dissolve by gentle heating in a fume head.

Oxidase Test Reagent

Solution A	
a-Naphthal	1 g
95% Ethanol	100 ml
Solution B	
Dimethyl-p-phenylene diamine Oxalate	1 g
Water	100 ml

Mix freshly prepared solution B with solution A in 1:1 ratio.

Spore Stain (hot)

Malachite Green	
Malachite green	5 g
Distilled water	100 ml
Safranine	
2.5% solution of safranine in 95% ethyl alcohol	10 ml
Distilled water	100 ml

Spore Stain (cold)

Crystal violet solution

Solution A	
Phenol	1 g
Distilled water	74 ml
Solution B	
Crystal violet 2 g	
Ethyl alcohol	26 ml

Mix solution A and B

APPENDIX 2

Microbiological Media:

The composition of some of the media commonly used is given below. Sterilisation of media is accomplished in an autoclave at 15 lb psi pressure for 15-20 minutes. Dehydrated media in powder form are also available that can be reconstituted as per instructions specified by the manufacturer.

The composition given below is in grams per liter.

Basal Salt Agar or Broth (pH 7.0)

Sodium diphosphate (0.5 M)	100.0 ml
Potassium dihydrogen phosphate (1.0 M)	100.0 ml
Distilled water	800.0 ml
Calcium chloride (0.1 M)	1.0 ml
Magnesium sulphate (1.0 M)	1.0 ml
Note: Swirl	
Ammonium sulphate	2.0
Agar	15.0

Shake well untill all the ingredients are dissolved. Cool and add 10 ml of 1% dextrose aseptically

Eosin- methylene Blue Agar (pH 7.2)

Peptone	10.0
Lactose	5.0
Dipotassium phosphate	2.0
Agar	13.5
Eosin Y	0.4
Methylene blue	0.065

Gel Diffusion Agar

Sodium barbital buffer	100.0 ml
Noble agar	0.8

Glucose Fermentation Broth (pH 6.8- 7.0)

Peptone	5
Beef Extract	3
Glucose	10

Add 1.0 ml Phenol Red as indicator

Glucose Peptone Water (pH 6.8 -7.0)

Peptone	5.0
Glucose	2.5
Dipotassium hydrogen phosphate	1.0

Glucose-Peptone Acid Agar (pH 4.0)

Potassium dihydrogen phosphate	0.1
Magnesium sulphate	0.5
Peptone	5.0
Glucose	10.0
Agar	15.0

Glucose Salt Broth (pH 7.2)

Dextrose	5.0
Sodium chloride	5.0
Magnesium sulphate	0.2
Ammonium dihydrogen phosphate	1.0
Dipotassium hydrogen phosphate	1.0

Glycerol Yeast Extract Agar Supplemented with Aureomycin (pH 7.0)

Glycerol	5.0 ml
Yeast extract	2.0
Dipotassium phosphate	1.0
Agar	15.0

Add aureomycin @ 10 μ g per ml to sterile, cooled and molten agar.

Inorganic Synthetic Broth (pH 7.2)

Sodium chloride	5.0
Magnesium sulphate	0.2
Ammonium dihydrogen phosphate	1.0
Dipotassium hydrogen phosphate	1.0

Lactose Fermentation Broth

Use 10 g lactose in place of glucose in Glucose Fermentation Broth

Litmus Milk (pH 6.8)

Skim milk powder	100 ml
Litmus	0.75 g

Autoclave at 12 lbs psi for 15 minutes.

MacConkey's Broth (single strength)

Peptone	20.0
Sodium chloride	5.0
Bile salt	5.0
Potassium phosphate	5.0
Lactose	10.0
Neutral Red (10% aqueous solution)	10 ml

Dissolve peptone, sodium chloride and bile salt in water by heating. Adjust the pH to 8.0 and boil for 20 minutes. Cool, filter and adjust the pH to 7.4. Add lactose and neutral red. Distribute in tubes and sterilize at 115°C for 15 minutes.

Double strength MacConkey's broth is prepared by doubling the quantities of all ingredients except distilled water.

Mannitol Salt Agar (pH 7.4)

Beef extract	1.0
Peptone	10.0
Sodium chloride	75.0
D- mannitol	10.0
Agar	15.0
Phenol red	0.025

Milk Agar (pH 7.2)

Skim milk powder	100.0
Peptone	5.0
Agar	15.0

Autoclave at 12 lbs psi for 15 min.

Minimal Agar (pH 7.0)

Solution A **(pH 7.0)**

Potassium dihydrogen phosphate	3.0
Disodium hydrogen phosphate	6.0
Ammonium chloride	2.0
Sodium chloride	5.0
Distilled water	800 ml

Solution B **(pH 7.0)**

Glucose	8.0
Magnesium sulphate	0.1
Agar	15.0
Distilled water	200 ml

Sterilise solution A & B separately and mix.

If required add thiamine to solution B @ 0.001 g prior to autoclaving. To combined sterile and molten medium add 50 mg sterile streptomycin before pouring into plates.

MR-VP Broth (pH 6.9)

Peptone	7.0
Dextrose	5.0

Mueller- Hilton Agar (pH 7.4)

Beef infusion	300 ml
Casamino acids	17.5
Starch	1.5
Agar	17.0

Nitrate Broth (pH 7.2)

Peptone	5.0
Beef Extract	3.0
Potassium nitrate	5.0

Nitrogen Free Mannitol Agar or Broth (pH 7.3)

Mannitol	15.0
Dipotassium hydrogen phosphate	0.5
Magnesium sulphate	0.2
Calcium sulphate	0.1
Sodium chloride	0.2
Calcium carbonate	5.0
Agar	15.0

Nutrient Broth (pH 7.2)

Peptone	5.0
Beef extract	3.0
Sodium chloride	8.0

Nutrient Gelatin (pH 6.8)

Peptone	5.0
Beef Extract	3.0
Gelatin	120.0

Peptone Broth (pH 7.2)

Peptone	4.0

Phenol Red Dextrose Broth (pH 7.3)

Trypticase	10.0
Dextrose	5.0
Sodium chloride	5.0
Phenol red	0.018

Autoclave at 12 lbs psi for 15 min.

Phenol Red Lactose Broth (pH 7.3)

Use lactose in place of dextrose in Phenol Red Dextrose broth

Sabouraud Agar (pH 5.6)

Peptone	10.0
Dextrose	40.0
Agar	15.0

If required add aureomycin @ 10 µg per ml to sterile, cooled and molten medium

Starch Agar (pH 7.0)

Peptone	5.0
Beaf extract	3.0
Starch (soluble)	2.0
Agar	15.0

Sucrose Fermentation Broth

Use 10 g sucrose in place of glucose in Glucose Fermentation Broth.

Triple sugar Iron Agar (pH 7.4)

Beaf extract	3.0
Yeast extract	3.0
Peptone	15.0
Protease peptone	5.0
Lactose	10.0
Saccharose	10.0
Dextrose	1.0
Ferrous sulphate	0.2
Sodium chloride	5.0
Sodium thiosulphate	0.3
Phenol red	0.024
Agar	12.0

Thioglycollate Fluid (pH 7.1)

Peptone	15.0
Yeast extract	5.0
Dextrose	5.0
L-Cystine	0.75
Thioglycollic acid	0.3 ml
Sodium chloride	2.5
Resazurin	0.001
Agar	0.75

Trypticase Nitrate Broth (pH 7.2)

Trypticase	20.0
Disodium phosphate	2.0
Dextrose	1.0
Agar	1.0
Potassium nitrate	1.0

Trypticase Soy Agar (pH 7.3)

Trypticase	15.0
Phytone	5.0
Sodium chloride	5.0
Agar	15.0

Tryptone Broth

Tryptone	10.0

Add glucose @ 1% to prepare Tryptone Glucose Broth.

Tryptone-Glucose Agar

Tryptone	10.0
Glucose	5.0
Agar	15.0

Adjust pH to 6.7 and sterilise

Tryptone Glucose Yeast Extract Agar (pH 7.0)

Tryptone	5.0
Yeast Extract	2.0
Glucose	2.0
Agar	15.0

Yeast Extract Broth (pH 7.0)

Peptone	5.0
Beef extract	3.0
Sodium chloride	5.0
Yeast extract	5.0

Yeast-Extract Mannitol Agar

Mannitol	10.0
Dipotassium hydrogen Phosphate	0.5
Sodium chloride	0.1
Magnesium sulphate	0.2
Calcium carbonate	3.0
Yeast extract	0.2
Agar	15.0

Yeast Extract Peptone Medium

Yeast extract	3.0
Peptone	0.6

Adjust pH to 7.2 and sterilise at 121°C for 15 min.

Tryptone Soft Agar

Tryptone	10.0
Potassium chloride (reagent)	5.0 ml
Agar	9.0

Urea Broth

Urea broth concentrate (Filter sterilised solution)	10.0 ml
Sterile distilled water	90.0 ml

Add urea broth concentrate aseptically to presterilised and cooled water. Dispense 3 ml into sterile test tubes under aseptic conditions.

APPENDIX 3

BUFFERS

When bacteria are grown in a medium containing sugars, they may produce acids either as intermediate or as end products resulting in the accumulation of acidic substances and decrease in pH. The effect of these acidic compounds can be neutralised by addition of buffering agents such as Dipotassium hydrogen phosphate or Potassium dihydrogen phosphate. Buffers are also used in the assay of enzymes.

Buffers are prepared by mixing a weak acid with its salt. In such a mixture a weak acid is only weakly dissociated while its salt with alkali metal is strongly dissociated. Thus, whereas 0.1-mol l^{-1} acetic acid is only 1.35% dissociated, 0.1 mol l^{-1} sodium acetate is 97% dissociated. If hydrogen ions are added to such a solution, they will react with the high concentration of salt anions to from non-ionized acids. This weak acid once formed does not tend to ionize appreciably and at the same time its ionization is opposed by the high concentration of anions present. Therefore, hydrogen ions though added yet do not effect the pH of the medium. Likewise A weak alkali and its salt can also be used to prepare a buffer.

The composition of some of the commonly used buffers is given below.

1. Citerate Buffer

Stock Solutions

A: 0.1 M solution of citric acid (21.01 g / l)

B: 0.1 M Sodium citerate (29.41 g $C_6H_5O_7Na_3 . 2H_2O$ /l); Use of salt with 5.5 H_2O is not recommended

Mix x ml of solution 'A' and y ml of Solution 'B'. Make volume to 100 ml

x	y	pH	x	y	pH
46.5	3.5	3.0	23.0	27.0	4.8
43.7	6.3	3.2	20.5	29.5	5.0
40.0	10.0	3.4	18.0	32.0	5.2
37.0	13.0	3.6	16.0	34.0	5.4
35.0	15.0	3.8	13.7	36.3	5.6
33.0	17.0	4.0	11.8	38.2	5.8
31.5	18.5	4.2	9.5	40.5	6.0
28.0	22.0	4.4	7.2	42.8	6.2
25.5	24.5	4.6			

R.D. Lille, 'Histopathologic Technique' Blakistan, Phiola delphia and Toranto (1948)

2. Acetae Buffer

Stock Solutions

A: 0.1 M solution of acetic acid (11.55 g / l)
B: 0.1 M Sodium acetate (16.4 g $C_2H_3O_2Na$ or 27.2 g of $C_2H_3O_2Na.3H_2O$ /l)

Mix x ml of solution 'A' and y ml of Solution 'B'. Make volume to 100 ml

x	y	pH	x	y	pH
46.3	3.7	3.6	20.0	30.0	4.8
44.0	6.0	3.8	14.8	35.2	5.0
41.0	9.0	4.0	10.5	39.5	5.2
36.8	13.2	4.2	8.8	41.2	5.4
30.5	19.5	4.4	4.8	45.2	5.6
22.5	4.5	4.6			

G. S. Walpole (1914), J. Chem. Soc. **105**: 2501

3. Tris (hydroxymethyl) aminomethane - maleate (Tris- maleate) Buffer

Stock Solutions

A: 0.2 M solution of Tris acid maleate [24.2 g of Tris (hydroxymethyl) aminomathane + 23.2 g of maleic acid or 19.6 g of maleic anhydride / l]
B: 0.2 M NaOH

Mix 50 ml of 'A' and x ml of 'B' and make volume to 200 ml

x	pH	x	pH
7.0	5.2	48.0	7.0
10.8	5.4	51.0	7.2
15.5	5.6	54.0	7.4
20.5	5.8	58.0	7.6
26.0	6.0	63.5	7.8
31.5	6.2	69.0	8.0
37.0	6.4	75.0	8.2
42.5	6.6	81.0	8.4
45.0	6.8	86.5	8.6

G. Gomori (1948), Proc. Soc. Expl. Biol. Med. **68**: 354

4. Tris (Hydroxymethyl) aminomethane - maleate (Tris-maleate) Buffer

Stock Solutions

A: 0.2 M solution of Tris (hydroxymethyl) aminomathane([24.2 g / l).
B: 0.2 M HCl

Mix 50 ml of 'A' and x ml of 'B' and make volume to 200 ml

x	pH	x	pH
5.0	9.0	26.8	8.0
8.1	8.8	32.5	7.8
12.2	8.6	38.4	7.6
16.5	8.4	41.4	7.4
21.9	8.2	44.2	7.2

5. Boric Acid – Borax Buffer

Stock Solutions

A: 0.2 M solution of boric acid (12.4 g/ l)
B: 0.05 M solution of borax (19.05 g / l ; 0.2 M in terms of sodium borate)

Mix 50 ml of solution 'A' and x ml of 'B' and make volume to 200 ml

x	pH	x	pH
2.0	7.6	22.5	8.7
3.1	7.8	30.0	8.8
4.9	8.0	42.5	8.9
7.3	8.2	59.0	9.0
11.5	8.4	83.0	9.1
17.5	8.6	115.0	9.2

W. Homes (1943), Anat. Record **86**: 163

6. Borax- NaOH Buffer

Stock Solutions

A: 0.05 M solution of borax (19.05 g / l ; 0.2 M in terms of sodium borate)
B: 0.2 M NaOH

Mix 50 ml of solution 'A' and x ml of 'B' and make volume to 200 ml

x	pH	x	pH
0.0	9.28	29.0	9.70
7.0	9.35	34.0	9.80
11.0	9.40	38.6	9.90
17.6	9.50	43.0	10.00
23.0	9.60	46.0	10.10

W. M. Clark and H. A. Lubs (1917), J. Bacteriol. **2:** 1

7. Phosphate Buffer

Stock Solutions

A: 0.2 M solution of monobasic sodium phosphate (27.8 g / l)
B: 0.2 M solution of dibasic sodium phosphate (53.65 g of $Na_2HPO_4.7H_2O$ or 71.7 g of 53.65 g of $Na_2HPO_4.12H_2O$ / l)

Take x ml of 'A' and y ml of 'B' and dilute to a total of 200 ml

x	y	pH	x	y	pH
93.5	6.5	5.7	45.0	55.0	6.9
92.0	8.0	5.8	39.0	61.0	7.0
90.0	10.0	5.9	33.0	67.0	7.1
87.7	12.3	6.0	28.0	72.0	7.2
85.0	15.0	6.1	23.0	77.0	7.3
81.5	18.5	6.2	19.0	81.0	7.4
77.5	22.5	6.3	16.0	84.0	7.5
73.5	26.5	6.4	13.0	87.0	7.6
68.5	31.5	6.5	10.5	90.5	7.7
62.5	37.5	6.6	8.5	91.5	7.8
56.5	43.5	6.7	7.0	93.0	7.9
51.0	49.0	6.8	5.3	94.7	8.0

S. P. L. Srensen (1909), Biochem. Z. **21:** 131; **22:** 352

8. Glycine- NaOH Buffer

Stock Solutions

A: 0.2 M solution of glycine (15.01 g / l)
B: 0.2 M NaOH

Mix 50 ml of solution 'A' and x ml of 'B' and make volume to 200 ml

x	pH	x	pH
4.0	8.6	2.4	9.6
6.0	8.8	27.2	9.8
8.8	9.0	32.0	10.0
12.0	9.2	38.6	10.4
16.8	9.4	45.5	10.6

S. P. L. Srensen (1909), Biochem. Z. **21:** 131; **22:** 352

9. Carbonate- Bicarbonate Buffer

Stock Solutions

A: 0.2 M solution of anhydrous sodium carbonate (21.2 g / l)
B: 0.2 M solution of sodium bicarbonate (16.8 g / l)

Mix x ml of solution 'A' and y ml of 'B' and make volume to 200 ml

x	y	pH	x	y	pH
4.0	46.0	9.2	27.5	27.5	10.0
7.5	42.5	9.3	30.0	20.0	10.1
9.5	40.5	9.4	33.0	17.0	10.2
13.0	37.0	9.5	35.5	14.5	10.3
16.0	34.0	9.6	38.5	11.5	10.4
19.5	30.5	9.7	40.5	9.5	10.5
22.0	28.0	9.8	42.5	7.5	10.6
25.0	25.0	9.9	45.0	5.0	10.7

G. E. Delory and E. J. King (1945), Biochem. J. **39:** 245

10. Succinate Buffer

Stock Solutions

A: 0.2 M solution of succinic acid (23.6 g / l)
B: 0.2 M NaOH

Mix 25 ml of solution 'A' and x ml of 'B' and make volume to 200 ml

x	pH	x	pH
7.5	3.8	26.7	5.0
10.0	4.0	30.3	5.2
13.3	4.2	34.2	5.4
16.7	4.4	37.5	5.6
20.0	4.6	40.7	5.8
23.5	4.8	43.5	6.0

NOTES

NOTES

NOTES

NOTES